Foreign Fronts of the French Air Force 1914–1919

David Méchin

Foreword

After having dealt with the intervention of the French air force on the Orient front, I thought of presenting in another volume all the other fronts of the First World War where the French air force was engaged, from the steppes of Russia to the sands of Africa, passing through the mountains of Romania and the lagoon of Venice. The reader will be able to find there all the information and statistics of these forgotten fronts presented for the first time to the English-speaking public.

David Méchin

Dedication

The author would like to thank his many friends for their help on this volume, and especially Paolo Varriale for his many exchanges on the Italian front, Boris Ciglic for his iconographical help on the Romanian front, and Pierre Jarrige who opened his personal archives on French aviation in north Africa. Special thanks to Robert Feuilloy and Lucien Morareau (association ARDHAN) for their help on naval aviation.

Acknowledgements

For our aviation books in print please see our website at: **www.aeronautbooks.com**

Thanks to Mike O'Neal for the cover painting.

For questions or to help with photographs please contact Aeronaut Books at **jherris@me.com**

ISBN: 978-1-953201-70-6

Cover Painting: Dragan Šaler
Design and layout: Jack Herris
Cover design: Aaron Weaver
Digital photo editing: Jack Herris

Aeronaut Books

Books for Enthusiasts by Enthusiasts
www.aeronautbooks.com

Table of Contents

Chapter 1: French Flyers in Italy 1915–1918

Above & Facing Page: The Nieuport NB2, taken out of its hangar and launched in the port of Brindisi from where it flew for a maritime patrol. (ARDHAN)

Officially a member of the Triple Alliance with Germany and Austria-Hungary, the Kingdom of Italy remained neutral when the First World War broke out in the summer of 1914. Not very sensitive to the appeals of the German government, Italy was much more sensitive to those of the British and French governments who urged it to enter the war on their side. Italy had a serious territorial dispute with the Austro-Hungarian Empire, which held under its control "irredent lands" populated by Italians, the Trentino region and the port city of Trieste.

Italy declared war on Austria-Hungary on May 23, 1915, and very quickly a front line was established... more or less on the borders. Indeed, the entire arc of the border between the two belligerents was located in the Dolomites and the Julian Alps, making any war of movement impossible. The only way to get through was the area to the east of the front, where, behind the Isonzo river, the Carso plateau was located on the coastal strip, making it possible to reach the city of Trieste, which was barely 20 km away.

For General Cardona, commander of the Italian army and a stubborn man, Trieste was the major objective. Until October 1917, he launched no less than 11 offensives, known as the Battles of the Isonzo, which all broke down on the Austrian trenches of the Carso, causing the loss of 300,000 Italian soldiers (half of the Italian losses of the war) against 200,000 Austro-Hungarians. In this war of pillboxes, the air force played an important role, as it did on the French front, in the adjustment of artillery. But the nearby Adriatic Sea brought a naval dimension to the war, with the use of fast boats (the Italians armed small, very fast ships with torpedoes, the MAS - Motoscafo Armato Silurante), pocket submarines or flat-bottomed monitors pounding the coastal part of the front with their artillery. Above the waves, seaplanes flew, used by the Austrians from their bases in Trieste or Pola and by the Italians in Venice. It was precisely in the field of naval aviation that France, which was to provide significant air support to its ally during the four years of the conflict, sent its first contingent of reinforcements.

NB.2

NB2

Above: View of the Nieuport coded "NB 3" of the Brindisi squadron, which carried out maritime patrols south of the Italian boot from September 1915 to January 1916. It was equipped with three Nieuport seaplanes and a prototype Caudron seaplane (Paolo Varriale).

1.1 – Air patrols over Brindisi

When it entered the war against Austria on May 23, 1915, the Kingdom of Italy was poorly equipped in terms of naval aviation, and its entry into the conflict was conditional on assistance in this area from its Franco-British allies. The French Navy then sent a small escadrille of four Nieuport seaplanes to defend this Italian port in the south of the peninsula.

The Nieuports, under the command of Lieutenant de Vaisseau Hautefeuille, were taken from the reinforcements that were to go to Port Said. Three of them (registered NB 1 to NB 3) disembarked on 14 September and their assembly began without delay.

Their performance was judged by Capitano di Fregata Ludovico De Filippi, commander of the *Elba* seaplane transport stationed in Brindisi, who wrote on September 16, 1915 to the head of the 2nd naval squadron: "These monoplanes develop a maximum speed of 120 km/h, and, provided that the sea is calm, can easily take off, reaching their patrol altitude of 2000 meters in 55 minutes (...) Their normal patrol time is three hours. The observer sits in front of the pilot, each aircraft is equipped to carry small bombs (...)

The three French pilots are naval non-commissioned officers, only the fourth, who is expected shortly, is an officer. The observers will have to be provided by the Regia Marina (...) These Nieuports are rather fragile and delicate machines, requiring perfect weather conditions to take off or land. Therefore, I don't think they are suitable for long range missions at sea; moreover, the only armament they carry is a simple automatic pistol (...) I suppose that the most judicious use of these aircraft would be to send them on occasional coastal patrols, just before the arrival or departure of a friendly ship in port (...) or to use them to search for enemy submarines in liaison with our own destroyers."

The first flights began on September 25 with Italian Navy observers trained on site, carrying two 105 mm shells for any offensive armament. They patrolled the Otranto Channel without encountering the enemy. On December 17, 1915, the squadron received the prototype of the Caudron "R" seaplane, a biplane with floats, which according to the documents was offered, then declined, by the Italian Navy to carry out torpedo dropping tests.

Above: The Caudron R, a unique aircraft that remained a prototype. The only floatplane of military aeronautics, this aircraft equipped with a 150 hp Canton-Unné engine remained in service until 1917 at the CAM Salonique.

The squadron was disbanded after 4 months of activity on January 17, 1916, when Italian naval aviation was deemed to be able to take over. The Nieuports and the Caudron, which had not suffered any losses, were assigned to Port Said for the former and Salonika for the latter.

1.2 – From coastal surveillance to the protection of Venice: a strong presence at the beginning of the war with CAM Venice and Escadrille N 92/I

This Brindisi squadron was preceded by another formation of six FBA 90 hp Gnome seaplanes (called type C by their manufacturer) which moved to Venice on June 2, 1915 to eventually form a centre d'aviation maritime (CAM, Maritime Air Center). Headed by Lieutenant de Vaisseau (captain) Antoine Reynaud, the center had six other pilots, including the famous Lieutenant de Vaisseau Jean Louis Conneau, who was well known in Italy for having won the Paris-Rome race on a Blériot XI on May 31, 1911 under the pseudonym André Beaumont. Incidentally, he was one of the shareholders of the company that built the seaplanes he flew and did not hesitate to act as a sales representative and to push for an order from the Italian Navy...

He soon had the opportunity to lead his aircraft into the fire, because if Brindisi was out of reach of the Austrian air force, this was not the case for the city of Venice, which was overflown on 24 May 1915 by two enemy seaplanes of the Lohner type, which had come from the naval air station of Trieste. The two bombs they dropped on the city did not cause much damage but caused a lot of excitement. Two other raids took place on May 27, 1915, and then on June 8, when the French seaplanes were in place on the islet of San Andréa, on the lagoon. Unfortunately, the pilots did not have any facilities on the island, were staying in Venice, and could not take off in time! LV Reynaud remedied this situation by organizing barracks and an alert network, which proved to be effective enough to allow an FBA to intercept a Lohner at dawn on July 4, 1915 and put it to flight. This success remained limited, as the Austrians returned with their seaplanes on 8 and 13 July 1915 and used the cloud cover to sneak into the city and bomb it without being bothered.

The French military attaché in Italy took up the matter and proposed to Paris that a land-based fighter unit be sent to protect the city that was dear to the hearts of the Italians. The general staff agreed and created the escadrille N 92 I (I for Italy), with three Nieuport 10 fighters commanded by Capitaine Marcel de Chalonge, which settled in Mestre, to the great displeasure of LV Reynaud, who saw this as a personal disavowal. Fate gave him the opportunity to take a

Profiles of the three Nieuport seaplanes of the Escadrille de Brindisi, all equipped with a 100 hp Clerget engine.

Cattaro
MONTE-
NEGRO
Antivari
(Bar)
Durazzo
(Durres)
Elbasan
ALBANIA
NB.2
Brindisi
Vlore
Otrante
Corfou

Above: In June 1915, a squadron of 6 FBA 100 hp seaplanes of the French Navy was installed on the island of San Andréa, which officially became the CAM Venise in 1916. The seaplanes were housed in a large hangar next to a channel, formerly used for torpedo tests. (Passot-Seguin collection)

Below: View of a Lohner seaplane in flight over the lagoons of the Italian coast. This aircraft, the only one used by the Austrian Navy in 1915 and early 1916, was the main adversary of French fighters and seaplanes. (Paolo Varriale)

Above: The Hansa Brandenburg C was the all-purpose aircraft of the Austro-Hungarian army, and was used throughout the war in constantly improved versions. Used on the land front of the Dolomite Alps, it was rarely encountered by the French pilots patrolling the Adriatic Sea, who found the seaplanes of the Navy in front of them. (Paolo Varriale)

Below: Symbol of the naval aviation war fought by the French pilots, this Austrian Lohner registered L 137. Hit by a ship's flak, it had to land on the Adriatic where it was captured by a fast torpedo boat of the Italian Navy, a MAS. (Paolo Varriale)

Above: Another view of Austrian Lohner registered L 137. (Paolo Varriale)

Left: Pre-war portrait of Ensign Jean Roulier, one of the three pilot officers of the CAM Venise at its creation. (Roulier family)

little personal revenge: on August 15, 1915, the very day the Nieuports arrived, the lookout network signaled the arrival of Austrian seaplanes on the Piave River. Three FBAs quickly took off and one of them, piloted by Enseigne de Vaisseau de 1ere classe (lieutenant) Jean Roulier with LV Jean Louis Conneau as observer, was able to intercept the L 49 Lohner of Linienschiffleutnant Heinrich Fontaine von Felsenbrunn, which dropped its bombs in the wilderness and turned around. He was pursued all along the Adriatic Sea by the French FBA, until the coast of Pola, where Conneau emptied the magazine of his Chauchat machine-gun without success.

The FBA seaplanes continued to try to intercept the enemy aircraft despite the presence of the Nieuport fighters. On 5 September, two Lohner seaplanes bombed Venice without being worried. During the night of October 24, four others bombed the railway station and returned at dawn, when only one of them had to turn back because of the interposition of an FBA. It was only from 18 November 1915

Above: Launching of a 100 hp FBA 100 hp (Type C) on the island of San Andréa, the aircraft coded "2" of Lieutenant Jean Louis Conneau, a famous pre-war civilian pilot known even in Italy for having won the 1911 Paris-Rome race on his Blériot (Passot-Seguin Collection)

Below: Ensign Jean Roulier, a young maritime pilot, poses in front of his personal FBA coded 10 in July 1915. This pilot became famous for his bombing attack on an Austrian submarine on July 1, 1915, the event making the front page of Italian newspapers (Passot-Seguin Collection))

Above: One might think that this is a photo of Lieutenant de vaisseau Jean Louis Conneau (in light-colored shirt and suspenders facing the camera) at the CAM Venise. But this is not the case: this is Lake Garda, whose northern end was occupied by Austrian troops. The Italian army hastily created a small seaplane squadron and bought FBAs from France. Conneau, an employee of the firm, delivered them in July 1915 (Passot-Seguin Collection)

Left: Chief of the Trieste naval air base throughout the war, Linienschiffleutnant Gottfried Banfield. Main opponent of the French aviators, he shot down three seaplanes of the CAM Venise on a total of 9 homologated aerial victories, making him the top ace of the Austrian navy. He was decorated by Emperor Franz Joseph shortly before his death, and then awarded the highest award of the empire by his successor, the Order of Maria Theresa, the only pilot to receive this honor (Paolo Varriale)

onwards that the Nieuports, whose number had increased to six, proved to be efficient: 5 Lohner seaplanes carried out a new raid that day. One of them, registered L 59, was almost shot down by the Nieuport piloted by Sous-Lieutenant (2nd lieutenant) Georges Lachmann (a future ace who finished the war with 9 victories) and returned to Trieste damaged. Installed from December 11, 1915 on an improvised airfield

Above: On August 15, 1915, four Nieuport 10s and as many pilots, plus the squadron leader of N 92 / I, Captain De Chalonge, set up on the Mestre airfield, on the shoreline west of Venice (SHD)

Above: The Nieuport 10s of the N 92 / I squadron were mostly used as single-seat fighters, but others were used as two-seaters. Their strength quickly increased to six aircraft in line in October. (Gallica)

Above: As early as September 1915, the French FBA 100 hp were equipped with a machine gun fixed on an improvised removable carriage. However, its limited range will limit its effectiveness in combat against Austrian aircraft. (ARDHAN)

Below: Nieuport 10 of escadrille N 92 / I flying over the Venice lagoon. The squadron left its Mestre airfield for the Lido on December 11, 1915 (SHD)

on the island of Lido, which was much more appropriate for cutting off the attackers, the Nieuports became an effective protection force for the city of Venice, so much so that the Austrians only ventured there at night, when the city was defended solely by anti-aircraft artillery.

The defense of Venice was not the only task of the French air force in Italy, contrary to what LV Reynaud believed. Shortly after his arrival, he was ordered to make himself available to the Italian Maritime Prefect. The latter asked him to use his seaplanes for reconnaissance over Trieste and the ports of Istria, using an advanced post that he set up in Porto Lignano and then on the coastal island of Grado. The reconnaissance missions were soon transformed into offensive missions, as they launched bombs at every opportunity, on objectives such as the Punta Salvore lighthouse, ships, and other military objectives. Thus, on July 1, 1915, EV1 Jean Roulier bombed the Austrian submarine U11 with his FBA, which earned him great fame throughout Italy, the exploit being reported in all the newspapers, even though the submarine left without a hitch... The escadrille N 92 / I escorted the FBAs in these missions on the Gulf of Trieste, detaching two of its Nieuports on October 19, 1915 to the advanced airfield of Casina Farello (on the shore near Grado). It also lent its observers and its Lewis machine guns to the CAM Venise, several missions being carried out in common.

If we are talking here about small raids carried out by a handful of French aircraft – two or three FBAs and as many Nieuport 10 two-seaters – we must not underestimate the importance of these raids, given the very small size of the Italian air force, which entered the war in 1915 with 14 squadrons of aircraft, many of which, such as Blériot XIs or Nieuport single-seaters, had very limited military value. In the fall of 1915, it had practically disappeared from the front... And it only maintained its presence thanks to the Voisin L and Farman XI delivered from France. The situation was very worrying, because if the Austrian Army aviation was hardly more numerous with a dozen squadrons posted on the Italian front, it included much more reliable aircraft, the Hansa Brandenburg type C (which would be used throughout the war with improved versions), used with competence and audacity on bombing raids on the back of the lines. Italian fighter aviation, non-existent in the summer of 1915, was only timidly launched in September 1915 at Udine with the commissioning of the first Nieuport 10s imported from France, forming the 8th Squadriglia (future

Above: Photograph of the Venice lagoon taken in autumn 1915 by Lt Louis Brullard, one of the pilots of the initial contingent of escadrille N 92 / I.

Below: On November 18, 1915, the first effective intervention of escadrille N 92 / I took place, when Second Lieutenant Georges Lachmann (on the right in the picture) succeeded in shooting down Lohner L 59, which he managed to damage. Lachmann left Venice in March 1916 to fight at Verdun. He finished the war with a total of 8 air victories, 5 of which were on the Russian front (SHD).

Above: View of Lohner L 59, which left Pola with 5 other similar aircraft to raid Venice. Its pilot, Linienschiffleutnant Demeter Konjovics, with Linienschiffleutnant Hugo Stenta as observer, returned to its base heavily damaged by the fire of Lachmann's Nieuport according to the Austrian archives (Paolo Varriale)

Right: On December 11, 1915, escadrille N 92 / I moved to the island of Lido, which was already a famous seaside resort at the time. Pilots and mechanics were housed in a requisitioned hotel, the Villa Paradiso. The comfort was quite different from that of the Adrian barracks that most of the pilots on the French front were familiar with... (SHD and Gregori Alegi)

On December 11, 1915, escadrille N 92 / I moved to the island of Lido, which was already a famous seaside resort at the time. Pilots and mechanics were housed in a requisitioned hotel, the Villa Paradiso, which is shown today. (SHD and Gregori Alegi)

Above: In January 1916, Gottfried Banfield recovered a seaplane designed for racing that he used as a single-seater fighter. Registered L 16, it would prove to be a formidable opponent to French and Italian aviation. (Paolo Varriale)

Above: Banfield's Lohner L 16 fighter, moored at its quay in Trieste (Paolo Varriale)

Above & Below: This picture, taken by an Austrian plane in May 1916, shows the installation of the two French units, on the mouth of the Venice lagoon. On the left, within the perimeter of the fort on the island of Lido, the airfield of escadrille N 392 (renumbered N 561 in June 1917). You can see the hangars but also the dangerous moats that border the airfield and in which more than one Nieuport will crash... Above, the island of San Andréa and its channel surrounded by hangars, housing the seaplanes of the CAM Venise and the Italian maritime aviation. (Mauro Antonelli)

Above: Another view of Fort San Niccolo on the Lido and the island of San Andréa (Paolo Varriale)

Right: The nationalist poet Gabrielle d'Annunzio, a regular visitor to the squadron, was photographed in June 1916 wearing the dark glasses he wore to recover from a shock following an air crash. Around him, French pilots of the N 392 squadron: from left to right, Lt René Robert, Lt Perron (observer), S/ Lt Darbos and Captain de Chalonge (SHD)

Above: View of the FBA of Ensign Vaugeois, lying on the quay in Trieste after his capture. Gottfried Banfield invited the French pilot to the table of his squadron where he and his pilots toasted his bravery. (Paolo Varriale)

Below: Ensign Jean Roulier examines his FBA 150 hp, just before his last flight on August 15, 1916, where he will be shot down by Banfield's fighter Lohner in the Gulf of Trieste (SHD)

Right: Wreck of the FBA 150 hp Hispano n°308 of ensign Jean Roulier, towed by the Austro-Hungarian sailors and brought back in Trieste. Only the body of Roulier was recovered by Italian sailors, while the observer, quartier-maître Auguste Costerousse, disappeared in the sea. (Paolo Varriale)

Below: Probably taken at the end of August 1916, this picture shows the inter-army cooperation of the French pilots of the CAM Venise and of escadrille N 392, often together in the air and on the ground. From left to right: Ensign Emery (observer), Chief Petty Officer Jules Duclos, Ensign André Woltz, Lieutenant de vaisseau Antoine Reynaud (head of CAM Venise), Second Lieutenant Gabriel Trouvé (of N 392), Medical Officer 1st Class Gabriel Espanet (dentist who became a Nieuport pilot before the war!), and Chief Petty Officer Paul Poggi (SHD)

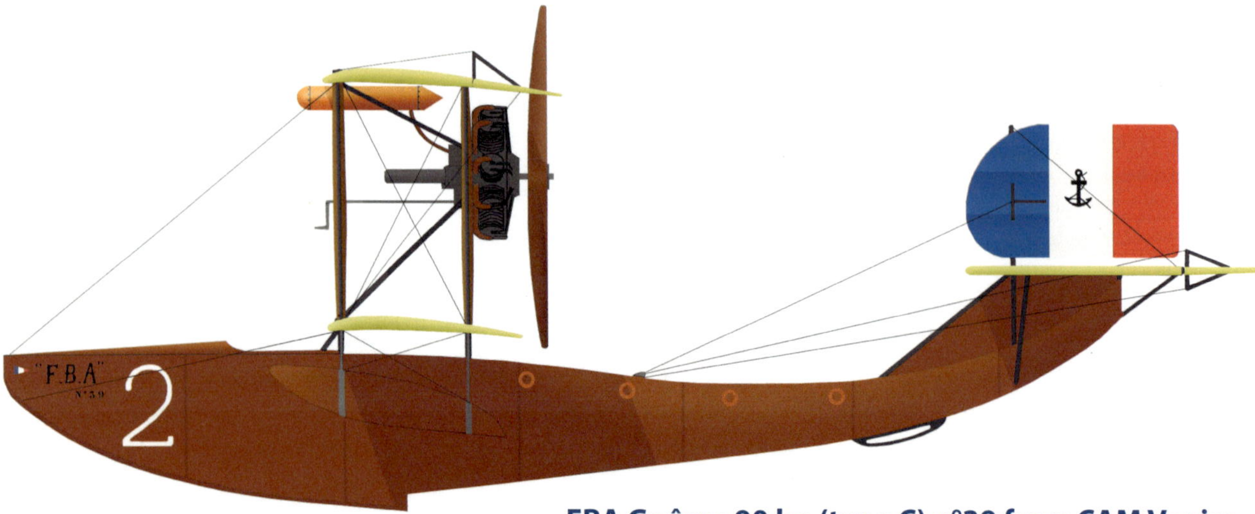

FBA Gnôme 90 hp (type C) n°39 from CAM Venise usually flown by lieutenant de vaisseau Jean Louis Conneau, San Andrea Island, Summer 1915.

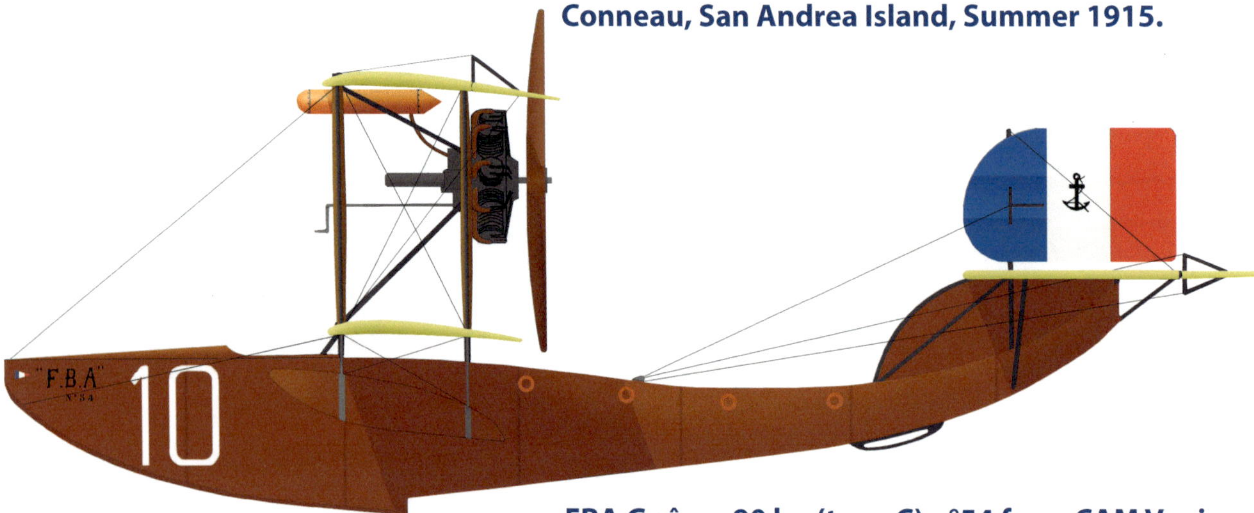

FBA Gnôme 90 hp (type C) n°54 from CAM Venise usually flown by enseigne de vaisseau Jean Roulier, San Andrea Island, Summer 1915.

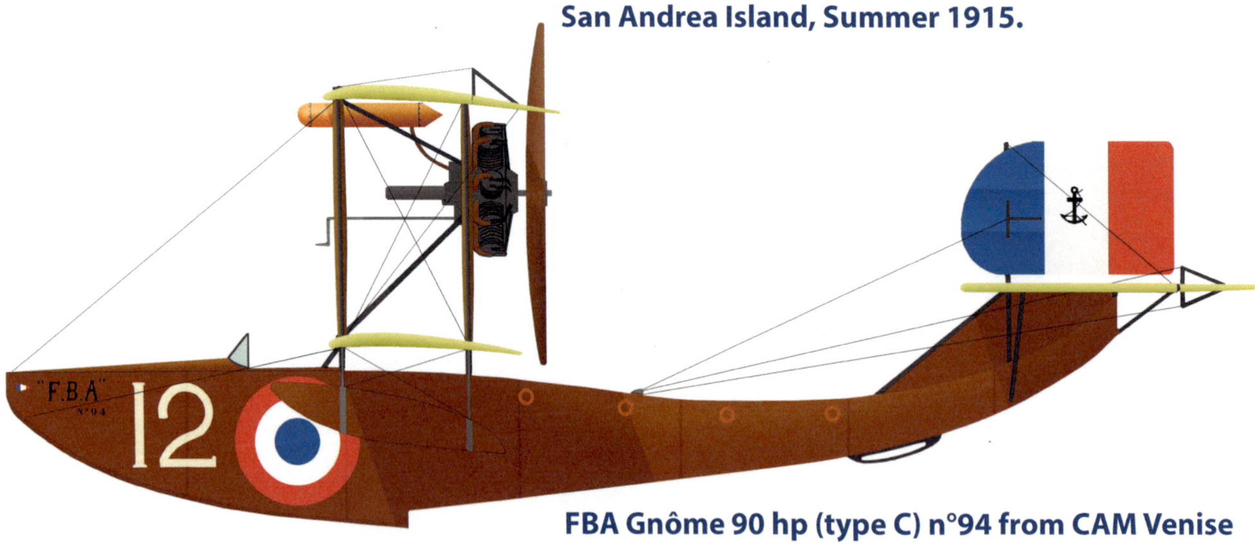

FBA Gnôme 90 hp (type C) n°94 from CAM Venise flown by enseigne de vaisseau André Vaugeois, San Andrea Island, June 1916. Vaugeois was captured in this plane.

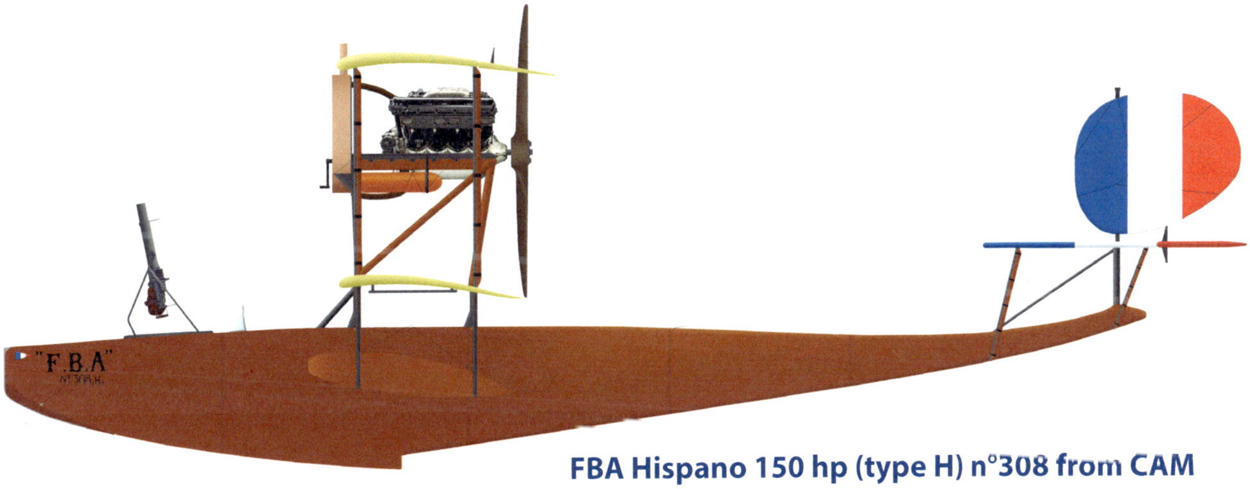

FBA Hispano 150 hp (type H) n°308 from CAM Venise flown by enseigne de vaisseau Jean Roulier, San Andrea Island, August 1916. Roulier was killed in this plane with his observer, quartier-maître Auguste Costerousse.

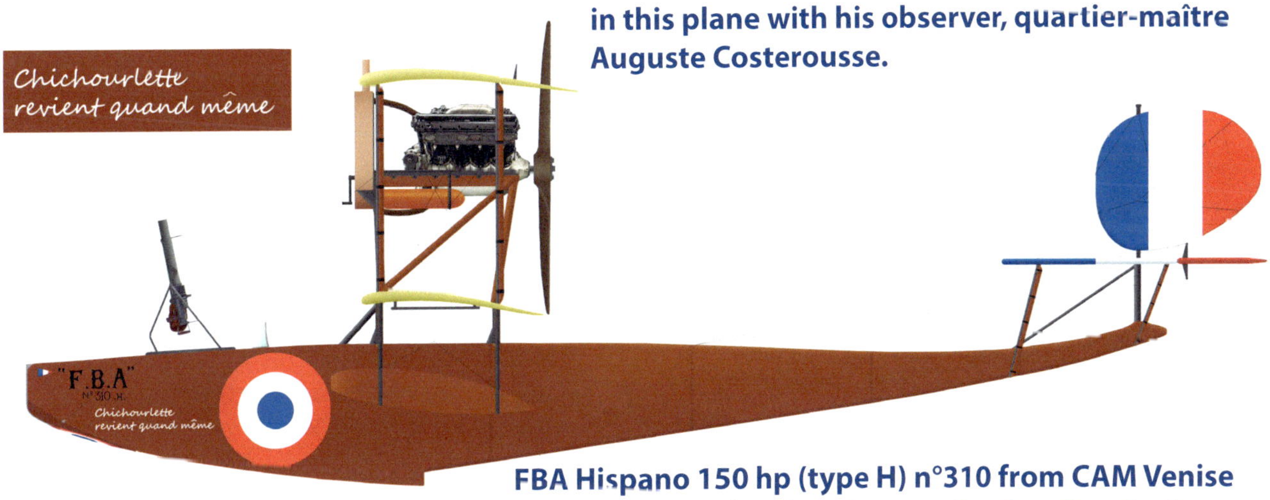

FBA Hispano 150 hp (type H) n°310 from CAM Venise flown by second maître Jacques Duclos, San Andrea Island, August 1916. Piloted by second maître Jacques Duclos with quartier-maître August Le Men as observer, this flying boat was shot down by Banfield but survived and was twed back to Venice. After repairs, it flew again with the nickname "*Chichourlette comes back anyway*".

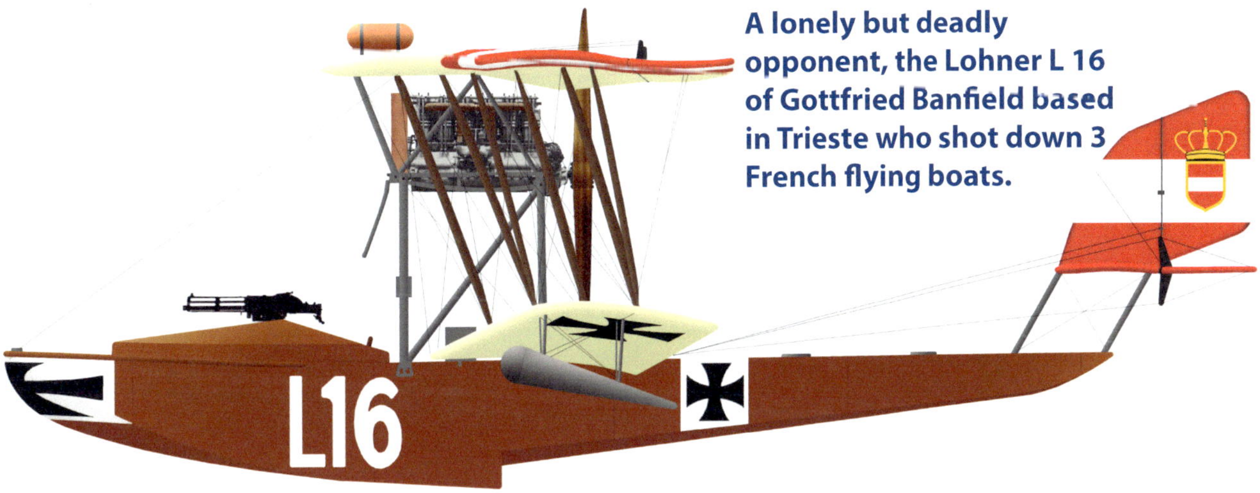

A lonely but deadly opponent, the Lohner L 16 of Gottfried Banfield based in Trieste who shot down 3 French flying boats.

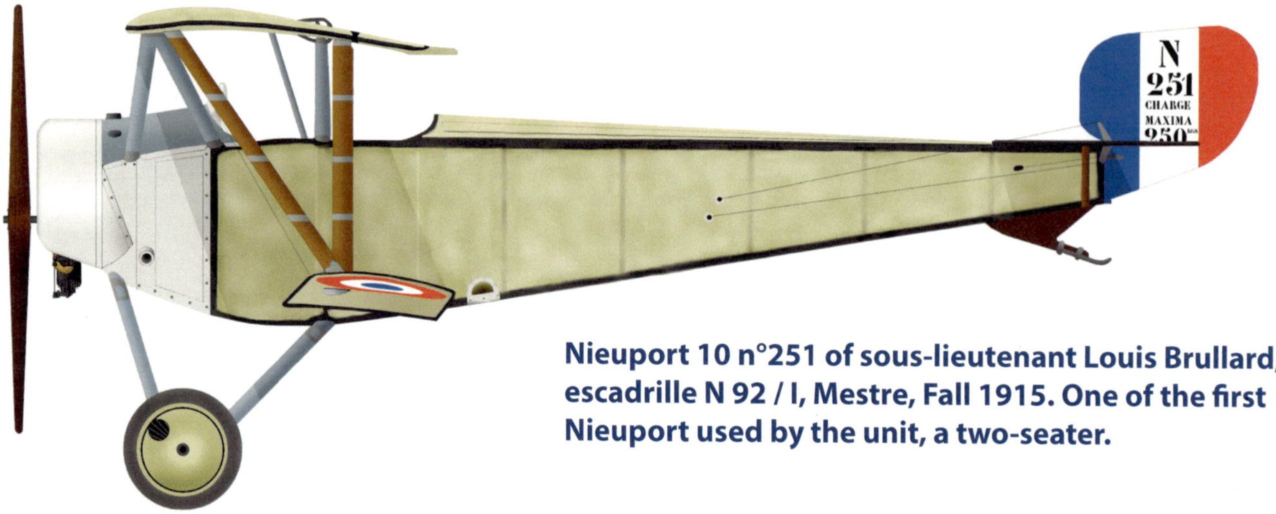

Nieuport 10 n°251 of sous-lieutenant Louis Brullard, escadrille N 92 / I, Mestre, Fall 1915. One of the first Nieuport used by the unit, a two-seater.

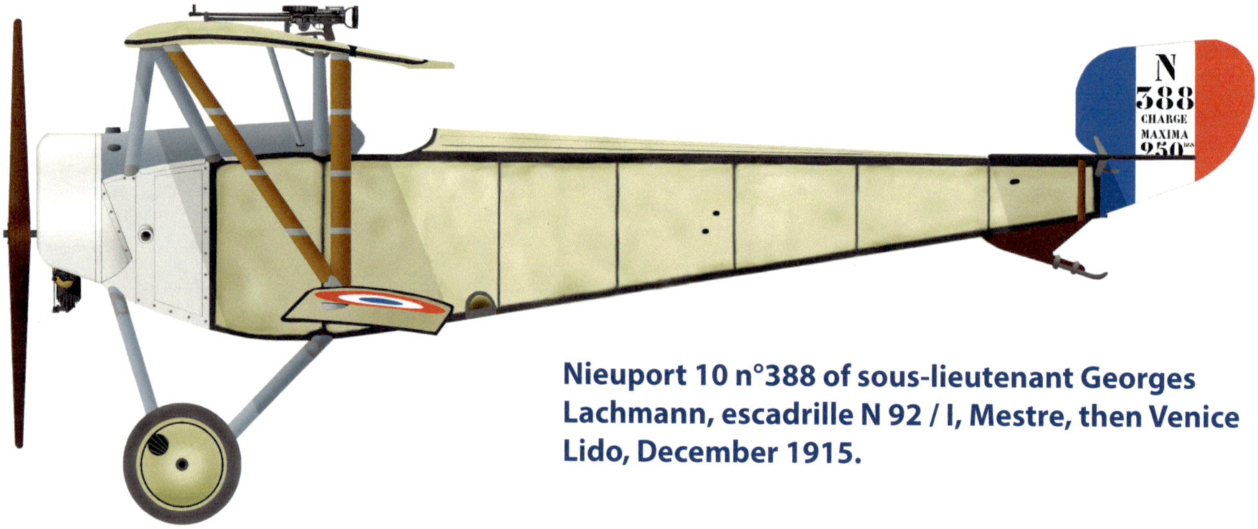

Nieuport 10 n°388 of sous-lieutenant Georges Lachmann, escadrille N 92 / I, Mestre, then Venice Lido, December 1915.

70th). The next units, the future 71st, 72nd, and 73rd, were only created in January 1916 at Cascina Farello, Brescia and Verona respectively, with the first Nieuport 11s. On the Adriatic, Italian maritime aviation was not much better off, and at the beginning of the war it only had two operational aircraft in Venice, two Albatros floatplanes. The numerical importance of the French air force in the summer of 1915 can therefore be better measured, as it concentrated its interventions on the Gulf of Trieste sector. Opposite, on the Austrian side, the seaplane squadron of the Trieste naval air base was present. When reading the reports written by the French, we realize that they were fighting a war against only one man: Linienschiffleutnant (navy captain) Gottfried Banfield, the leader of this unit.

The first shots were fired on August 7, 1915, when two French FBAs flew over Trieste (Roulier's "10" and Perron's "12"), escorting the two Italian Albatros seaplanes piloted by the tenente di vascello (TV - navy captain) Giuseppe Miraglia, head of the Italian naval aviation, and TV Luigi Bologna. It was a mission to drop leaflets for the indigenous Italian population – leaflets written by the poet Gabriele d'Annunzio who was in one of the Albatroses as an observer. The four aircraft were intercepted over the city by two Lohners, which were countered by the French, each of them being shot at and making a few holes in the canvas, without any major consequences. One of the Austrian seaplanes was piloted by Banfield who met the French in the Gulf of Trieste for the first time.

It would not be the last. On September 1, 1915, an epic battle took place, in which Banfield reported having faced no less than three French seaplanes. The French navy archives, which can be consulted at the Château de Vincennes, provide details of the battle. The wreck of an unknown submarine was reported 4 miles south of Sdobba Point. The CAM Venise was ordered to recognize the nationality of the minefield that sank it. Two FBAs took off from Grado at 14:30, the "2" piloted by Lieutenant de Vaisseau Conneau, taking as observer the Tenente di Vascello Giambernardini, deputy commander of the Grado station, accompanied by the FBA "10" piloted by EV1 Roulier, taking as observer Lieutenant Romeyer, "loaned" to the CAM Venise by escadrille N 92 / I, and who was operating a Lewis gun taken from a Nieuport fighter that was fixed on a mobile mount. Roulier placed his FBA on the water at Punta Primero ready to attack any possible enemy. Meanwhile, Conneau's FBA "2" flew over the target for an hour and a half, spotting mines and landing on the water to examine them and inspect the wreckage of the submarine. As he was about to take off for Grado, an Italian MAS commanded by Tenente di Vascello Luigi Rizzo, a future hero of the Italian navy who sank no less than two major Austrian battleships, reported enemy aircraft approaching. He saw "a land plane at 3,000 meters and a seaplane dropping a bomb on the wreckage of the submarine before attacking the MAS". The seaplane was the Lohner L 46 piloted by Banfield. Conneau flew over Roulier to warn him. Roulier took off and gained altitude, but Banfield saw him and fell on him. Romeyer fired two magazines but the fixation of his Lewis gun did not allow him to shoot sufficiently upwards. On the other hand, Banfield's machine-gunner named Strobl hit him with 6 explosive bullets that did only minor damage.

Banfield, believing he had finished with Roulier, charged at Conneau, but the latter flew to the Grado lagoon under cover of the Italian flak, which fired at the Lohner, forcing it to move away. But Giambernardini, who was only on his second flight, could not stand this aerial ballet and seemed to be suffering from airsickness, giving "some signs of weakness"... Conneau landed in the port and asked for a volunteer to replace him. The French sailors remained silent, much to his dismay. But one of them, Matelot (Seaman) Rouanet, was not afraid to take on the task and climbed into the FBA, grabbing the machine gun. After refueling, Conneau immediately took off and found Banfield, who ran into him before he reached his height. Conneau passed him at a distance of 300 meters and 200 meters below, then placed himself in the blind spot of the Lohner's machine gun. Not being able to shoot because of the distance, he returned to the lagoon of Grado to gain altitude until 1700 m while being protected by its Flak, then he returned a third time against Banfield, who broke the combat by diving towards Trieste, accompanied by a burst of Chauchat that Rouanet shoots from too far to touch him. Down to 900 m, the French pilot sees him landing near 2 Austrian patrol boats. He returns on his side to Grado after a fight which lasted more than one hour!

Roulier then set out again in a fighter patrol at 16h40 on the FBA "10" but did not find anybody any more. Banfield indicated in his report that he had confronted no less than four enemy seaplanes, one of them twice. He probably mistook the attacks of the relentless Conneau for those of several aircraft. In one of the curious ironies of which history is sometimes capable, this was the second time the two men crossed paths. The first one took place in 1913 in Argenteuil, where Conneau, chief pilot at Donnet-Levèque company, gave his first flying lessons to Banfield... The master and the pupil did not meet again since Conneau was recalled to France at the end of September 1915.

The winter would limit the number of flights, but the fighting would continue in 1916. While the Italian naval aviation was developing rapidly, receiving at the end of 1915 Macchi L.1 seaplanes copied from the Austrian Lohner, the CAM of Venice suffered from the wear and tear of its equipment. The seaplanes received to replace its FBA 100 hp, Donnet-Dennaut 150 hp, were of very poor quality and were only used for local flights over Venice, reserving for the FBAs the missions over the Gulf of Trieste. Escadrille N 92 / I, renumbered N 392, received an abundance of Nieuport 10 single-seaters and two-seaters and saw its number of pilots increase to 8. Opposite, Gottfried Banfield flew solo on a special Lohner seaplane fighter armed with a fixed machine gun that he used for air combat. It was this aircraft, registered L16, that Enseigne de Vaisseau André Vaugeois, a new pilot of CAM Venise, met on June 23, 1916. During a shooting pass Banfield killed his observer, an Italian from Austria-Hungary named Grammaticopoulo, and damaged the FBA which had to land on the water. Vaugeois, after a courageous fight with machine gun fire and an attempt to reach his lines by sailing, was finally captured by the Austrians. The French archives reveal that Banfield had one of his seaplanes drop a message in Venice informing the French of the outcome of the battle, mentioning the heroic resistance of the pilot. He also let pass two French seaplanes that took off from Grado to look for their comrade and could clearly guess what happened to him.

For the French as well as the Italians, the L16 was the plane to shoot down. Three days later, on June 26, 1916, escadrille N 392 detached again some aircraft to Cascina Farello and organized a trap against Banfield. A Macchi L 1 flying close to the shore, with a well-cleaned tank to be seen from far away, was used as bait. When Banfield flew to intercept it, the Italian turned back towards the open sea: fast boats in which Captain de Chalonge was seated set up a flak barrage, and several Nieuport 10s fell on him, one piloted by Lieutenant Daniel Chambarière, the other one being a two-seater piloted by Lieutenant René Robert and taking Lieutenant Romeyer as observer. Banfield managed to escape, pulled by the two Nieuports, while other Austrian seaplanes that came to his rescue were countered by a third Nieuport flown by Doctor Gabriel Espanet, one of the test pilots of the Nieuport company. The operation, if it did not shot down its target, was considered a success and was celebrated with a feast in the squadron's mess hall, which Gabriele d'Annunzio honored with his presence by dedicating the menu with the word "higher!"

1.3 - The French withdrawal (summer 1916–October 1917)

By the end of June 1916, the Italian army fighter force had seriously developed. Equipped with Nieuport 11s manufactured under local license, it was able to benefit from French flying schools that trained a first contingent of pilots, the next ones of which began to emerge from those set up in Italy. 9 fighter squadrons of the Corpo Aeronautico Militare

Left & Above: First confirmed aerial victory of escadrille N 392, the Lohner L 138 shot down by Sous-Lieutenant Xavier Garros and Lieutenant René Robert. The Austrian seaplane landed in a swamp near Caorle, north-east of Venice, where its crew, composed of the Seekadet Frantz Viranyi and the observer Hoch, was captured by the soldiers of the guardia di finanza (Paolo Varriale)

Above: A crater, a burnt piece of hull and a tattered hangar, that's what we find on this picture taken after October 31, 1916 when the accidental explosion of a bomb that was being loaded on the FBA 150 hp n°321 killed LV Reynaud (CAM Venise's commanding officer), EV Emile Hariat (observer), 13 French and 3 Italian sailors, plus 8 French and 9 Italian wounded. The CAM Venise was decimated and began its long decline... (Paolo Varriale)

Above: After the accident of October 31, the CAM Venize had plenty of FBA 150 hp Hispano, which can be seen here lined up along the San Andrea channel. But there were only four pilots left to fly them, in addition to its new commander and sole officer, Enseigne de vaisseau Woltz, who himself was killed in action on April 17, 1917 (Paolo Varriale).

Above: Entered in the inventory in September 1916, FBA n°329 coded H7 is the usual mount of second maître Eugene le Guennec, who is about to take off from the San Andréa channel (Paolo Varriale)

Below: Decoration ceremony for French pilots in Piazza San Marco, Venice, during the winter of 1916-1917. From left to right: Lieutenant René Robert, enseigne de vaisseau André Woltz, sous-lieutenant Xavier Garros, second maître Paul Poggi, maréchal des logis Edouard Corniglion, plus an unidentified sailor. Robert, Garros and Corniglion took part in the air battle that saw the capture of the L 138 seaplane on October 23, 1916, the first certified victory of escadrille N 392. (SHD)

Above: On December 7, 1916, Adjudant André Mélin completely missed his landing and destroyed his Nieuport 11 in the lagoon. The pilot would be sent to hospital. (SHD)

Below: In May 1917, the French army sent to Italy a detachment of heavy artillery on railroads (ALVF) which went into action on the Trentino front. Along with the long-range guns, 6 Sopwith 1A2s were sent for artillery spotting, grouped together in a special unit, escadrille Espinasse. (Paolo Varriale)

Above: The officer-pilots of escadrille Espinasse are received by the pilots of escadrille N 561 at Villa Paradiso. In the center of the picture, in light pants, capitaine de Chalonge, head of N 561. Immediately to the right, capitaine Jean de Fontenilliat, grand-nephew of General Espinasse, a hero of the Italian War of Independence who died in 1859 at the Battle of Magenta. Among its pilots, a personality: sous-lieutenant Paul-Louis Weiller (second from right), hero of the French reconnaissance aviation and captain of industry during the interwar period (SHD).

Left: Integrating N 561 on July 7, 1917 in a contingent of four pilots, Sergeant André Lévy decorated his personal Nieuport with a bulldog, a decoration that was repeated on his following aircraft. (SHD)

A professional officer from the cavalry, as illustrated by the pilot's personal badge, aspirant Marcel Robert began his career with eight days' detention in a fortress... For refusing to be assigned to N 561, preferring to fight on the French front. He arrived in Italy on June 7, 1917, and left on his first leave in France in April 1918, ending the war at escadrille SPA 124. He won three air victories during the war, including one in Italy. He then had a long career in the French air force (SHD).

Left: Caricature of members of escadrille N 561. On the boat, capitaine de Chalonge, commander of the unit. On the left, S/Lt Jean Jacques Darbos. In the air, Lt Olphe Galliard, observer. On the left, Lt de Limur, a future famous film director of the interwar period (SHD)

Left: For pilots who are not on immediate alert, the island of Lido offers some distractions... Sous-lieutenant Jean Bignon and lieutenant Jean de Limur relax on the beach in deck chairs. One must admit that during the First World War, it was difficult to find a better assignment! (SHD)

Above & Below: Originality of the Italian front, the naval support of the troops on the ground. During the 11th battle of the Isonzo, the Italians used the "Faa di Bruno" monitor, a huge floating casemate built around two enormous cannons of an unfinished line ship. Its fire was regulated by Italian seaplanes. (Paolo Varriale)

Above: View of N 561's pilots posing around the Italian maritime prefect (center in white uniform), during the summer of 1917. From left to right: sergent Maurice Faure, sergent Etienne Robert, maréchal des logis Jean Tétard de Morcellier (shot down by enemy flak on September 4, 1917), maréchal des logis Edouard Corniglion (2 victories), lieutenant Frederic Loiseau, sergent Ivan Marcou, capitaine Marcel de Chalonge, maréchal des logis Maurice de Geoffre de Chabrignac (1 victory), sous-lieutenant Xavier Garros (2 victories), sous-lieutenant Jean Bignon (1 victory), sous-lieutenant Jean Jacques Darbos, and lieutenant Louis Olphe-Galliard (observer, 1 victory won at escadrille Espinasse) (SHD)

Above: View of the hangars of the Lido airfield, base of escadrille N 561, at the end of the summer of 1917. One can see three Nieuport 23s on alert: lieutenant Frédéric Loiseau's n°3361 "Elefante Néro", maréchal des logis Edouard Corniglion's n°3363 "Ocio - Fiol d'un can", and sergent Ivan Marcou's n°2100 "Rondinella". At this date, all the aircraft of the squadron carry a personal decoration. (Paolo Varriale)

Above & Below: Close-up view of Nieuport 23 n°3363 of MdL Edouard Corniglion. Very fluent in Italian language, he decorated his aircraft with the Italian imprecation "Ocio! Fiol d'un can", which translates as "Beware, son of a dog" (Paolo Varriale)

Above: Views of Lieutenant Frederic Loiseau's Nieuport 23 #3361 flying over the city of Venice. His Nieuport was christened "Elefante néro", or "Angry Elephant". (SHD)

Above: After the departure of escadrille Espinasse from Italy, escadrille N 561 recovered some of its Sopwith 1A2s and one of its observers, lieutenant Olphe-Galliard, seen here preparing for the flight. On September 4, 1917, this observer carried out a reconnaissance mission over Istria, flying over the port of Fiume. (Paolo Varriale)

Above: Another pilot of ecadrille N 561 specialized in the piloting of two-seaters, Lieutenant René Robert (1 victory), in front of one of the Sopwith 1A2 two-seaters of the squadron. He was recalled to France at the end of 1917 to take command of escadrille SOP 106. (Paolo Varriale).

Above: Nieuport 23 of Loiseau in flight.

were now operational at the beginning of the summer, reducing the numerical importance of escadrille N 392. As for the Italian navy aviation, already equipped with Macchi L 1 seaplanes, it received more powerful L 2s delivered in May 1916, in addition to FBA seaplanes ordered from France.

The CAM Venice, already suffering from the unavailability of its seaplanes, was also going to suffer from a shortage of pilots whose departures and losses were not replaced. In his correspondence, Ensign Roulier underlines the shame he felt in front of his Italian comrades by not being able to be of much use on his seaplanes when they were out of breath. He also notes that the Austrians are more and more efficient with their fighter seaplane piloted by Banfield... On August 15, 1916, a raid was launched on Trieste, in which 5 Macchi L 1 escorted by 4 Nieuports of escadrille N 392 participate. The CAM of Venice could only put in the air its 2 new FBA 150 hp Hispano (type H) that had just been delivered. Banfield managed to avoid the escort and shot down the 2 FBA. If one survived (FBA n°310 piloted by second maître Jacques Duclos with quartier-maître François Le Men) and could be towed to Grado, the other one piloted by Roulier with quartier-maître Auguste Costerousse as gunner was not so lucky and exploded in flight, killing the crew. Only Roulier's body was recovered by an Italian patrol boat and brought back to Venice where he was buried after a grandiose ceremony in which d'Annunzio read his eulogy. The day after this hard blow for the CAM Venice, a new officer arrived, Enseigne de Vaisseau de 1ere classe André Woltz, who discovered a squadron without any aircraft... On September 13, 1916, while the Italians launched a raid on Parenzo with 11 Macchi L 1s in retaliation for a night bombing of Venice, the CAM could only put in the air a single FBA 150 hp, piloted by Woltz... Nevertheless, the raid was the occasion for escadrille N 392, which ensured the escort of the seaplanes, to win its first aerial victory (Lt Chambarière) by forcing an enemy seaplane to ditch. The victory was not certified, but the Austrian seaplane, registration number K 163, was indeed lost – not from French bullets, but from an accident while hoisting it ashore for repairs.

A few days later, on September 21, 1916, CAM Venise finally received a batch of FBAs 150 hp Hispanos (type H) which solved its equipment problems, although the pilots were still lacking – there were only four of them left. Woltz found himself at the head of the CAM on October 31, 1916, when a bomb accidentally exploded on the ground and killed more than fifteen sailors, including LV Reynaud, commander of the center, not to mention the many injured. On April 17, 1917 Woltz himself disappeared in the fog off Venice while attacking an Austrian ship with his seaplane sounding the death knell for the French formation which no longer had any officers. In spite of the energy of Petty Officer Cadio, the oldest in the highest rank, who tried to keep the unit operational, it was disbanded on 27 June 1917. No Italian officer came to witness its departure, which surprised the administrative officer who came to liquidate it. The Maritime Prefect replied that "France, which was older than Italy, was a bit of a mother-in-law to her, and you know that in families we try to separate mothers-in-law from daughters-in-law!"

However, maritime aviation did not disappear from the Italian sky for long, because in Marsala, on the western tip of Sicily, an autonomous combat post was opened, depending on the CAM of Bizerte. Three FBA 150 hp Hispano seaplanes were permanently based there and participated in anti-submarine patrols between Sicily and Tunisia (see chapter 5).

As for escadrille N 392, renumbered N 561 in June 1917, only its pilots temporarily assigned to Cascina Farello were active due to the empty sky over Venice. In October 1916, the squadron consisted of 10 aircraft, Nieuport 11s and two Type 16s equipped with Le Prieur rockets, which were sent to the advanced airfield to carry out a Drachen hunt with the Italian pilots. A diplomatic incident occurred on October 18, 1916 when Commander Pier Ruggiero Piccio, head of the local Italian fighter squadron, took off in the morning without the French pilots, who strangely enough had not been picked up at the Grado landing stage, leaving them to reach Cascina Farello on foot... Piccio came back victorious from his solo Drachen hunt, winning the first of his 24 official aerial victories, but a serious quarrel with the French since Captain de Chalonge ordered his men to return to Venice! On October 23, 1916, the first confirmed victory of the squadron was won during an escort of Macchi L 1, at the expense of Lohner L 138 which had to land at Caorle where it was captured. This was the only significant event for the French pilots, who enthusiastically welcomed Captain Louis Robert-de-Beauchamp who landed near Venice on November 17, 1916 on his Sopwith 1½ Strutter (called Sopwith 1B1 by the French) after a daring solo bombing raid on Munich, taking off from Luxeuil in the east of France. Captain de Chalonge wanted his men to follow in his footsteps, and on November 20, he asked for Sopwiths to be delivered to him to bomb Vienna. The request remained unanswered and his pilots were left to remain in their role as sentinels of Venice.

Life was not so difficult for the pilots who experienced a certain dolce vita in their assignment. Several pilots have left picturesque testimonies of their activity in Italy, including Lieutenant Frédéric Loiseau who wrote "In the lottery as in war, there are good numbers...". The pilots were housed on the island of Lido, which at that time was already a seaside resort for wealthy tourists. A hotel, aptly named "Villa Paradiso", was requisitioned for them. Because of the diplomatic role of the squadron, the military commissariat did a good job of sending them talented cooks who made the squadron's kitchen a famous table. During the day, the pilots were divided into three groups: those on immediate alert, who had to be on the airfield, ready to take off for alerts that rarely came. Those on duty must remain at Villa Paradiso, ready to be called for reinforcements. As for the others, they had free time on the island of Lido, but could go to Venice with special authorization from the squadron leader. To kill time during the day, the pilots built a clay tennis court, which they sprayed

Above: Present in Italy in the fall of 1917, before the defeat at Caporetto, SOP 36 wore a frog on its Sopwith 1A2s, often associated with a personal badge. (Paolo Varriale)

Above: This Sopwith of SOP 36 wears a seahorse on the tail fin as a personal insignia. (Albin Denis)

Above: Mystery photo of these Sopwith 1A2s decorated with an unidentified insignia by the Air Force Historical Service, a magician perched on a smiling crescent moon and observing the ground with a long sight. One could think of escadrille Espinasse, but the serial numbers of the aircraft in the photo do not correspond to those attributed to it. It could therefore be a first abandoned badge of one of the French Sopwith squadrons present on the Italian front before the battle of Caporetto, SOP 208 or SOP 221 (Luigino Caliaro)

Above: Lieutenant Daniel Chambarière in his Nieuport 23 decorated with a very popular Italian cartoon character, the "cittadino che protesta" (the citizen who protests) that appeared on a humoristic magazine with the stories written by Oronzo Marginati, nom de plume of the writer Luigi Locatelli. "Ferma!!" means "Stop!!" in Italian. Chambarière ended the war as commanding officier of escadrille SPA 162 where he won 4 victories, having under his orders maréchal des logis Corniglion, as him veteran from escadrille N 561.

with gasoline from the plane to kill the earthworms that would damage the surface - a fact that was revealed by a mechanic to the French tabloid press, a mechanic who was promptly dismissed by Captain de Chalonge. In the evenings, when the squadron was not flying, it was free time for everyone – the most aristocratic pilots frequented good society and were easily received in the salons of the Venetian aristocracy. Many had bachelor apartments in town...

However, the dolce vita was not to everyone's taste, as they felt they were wasting their time while the fate of France was being decided at Verdun. Two pilots left their testimony to the historical service of the Air Force. Sergeant André Dubonnet, heir to the famous brand of alcohols and future racing driver, used his connections to be transferred to France and left the unit on August 19, 1917, four months after his arrival. This is also what Aspirant (officer cadet) Marcel Robert did on his first leave in France in April 1918. His exact words about N 561 was "A squadron with no military interest ... A first-class hideout! As soon as we appeared, the Austrians would get the hell out of there."

At the beginning of 1917, the squadron had 12 pilots and was equipped with Nieuport 17 or 23s that the pilots, who had time on their hands, decorated with splendid personal markings. A rare alert on April 17 allowed them to see some action and to shoot down a seaplane off Venice. In May, another unit arrived in Italy, the escadrille Espinasse, named after the French general Charles Marie Esprit Espinasse who died in 1859 during the battle of Magenta, which saw the victory of Franco-Sardinian troops against the Austrians. This general was also the great-uncle of the squadron leader, Captain Jean de Fontenilliat, who has under his command 6 Sopwith 1A2s intended to adjust the fire of a French Heavy Artillery detachment sent as reinforcement. The Nieuports of the Venice squadron provided him with a fighter escort

Facing Page, Below: The Lido airfield is located in the enclosure of the fort guarding the entrance to the Venice lagoon. In its moat, located all around, sometimes fell clumsy pilots who missed their landing, as it is the case of sergent Henri Boyer who crashed his Nieuport 23 decorated with a Pinocchio (SHD).

Nieuport 11 n°1404 of Adjudant André Melin, escadrille N 392, Cascina Farello, Oct 1916.

Nieuport 21 n°1953 of Maréchal des Logis Edouard Corniglion (2 victories), escadrille N 392, Venice-Lido, 1916-1917.

Nieuport 23 n°2066 of Sous-Lieutenant Pierre de Jumilhac, N 392, Venice-Lido, 1916-1917.

Nieuport 23 n°2100 of Sergent Ivan Marcou, N 392, Venice-Lido, 1917.

Nieuport 12bis n°2127 of Capitaine René Robert, N 392, Venice-Lido, 1917.

Nieuport 23 n°2750 of Sergent André Lévy (6 victories), N 561, Venice-Lido, 1917.

Nieuport 23 n°2756 of Sous-Lieutenant Jean Bignon (1 victory), N 561, Venice-Lido, 1917.

Nieuport 23 n°2903 of Aspirant Marcel Robert (3 victories, including 1 in Venice), N 561, Venice-Lido, 1917.

Nieuport 23 n°3341 of Lieutenant Frédéric Loiseau, N 561, Venice-Lido 1917.

Above & Below:
Nieuport 23 n°3363 of Maréchal des Logis Edouard Corniglion (2 victories), N 561, Lido 1917.

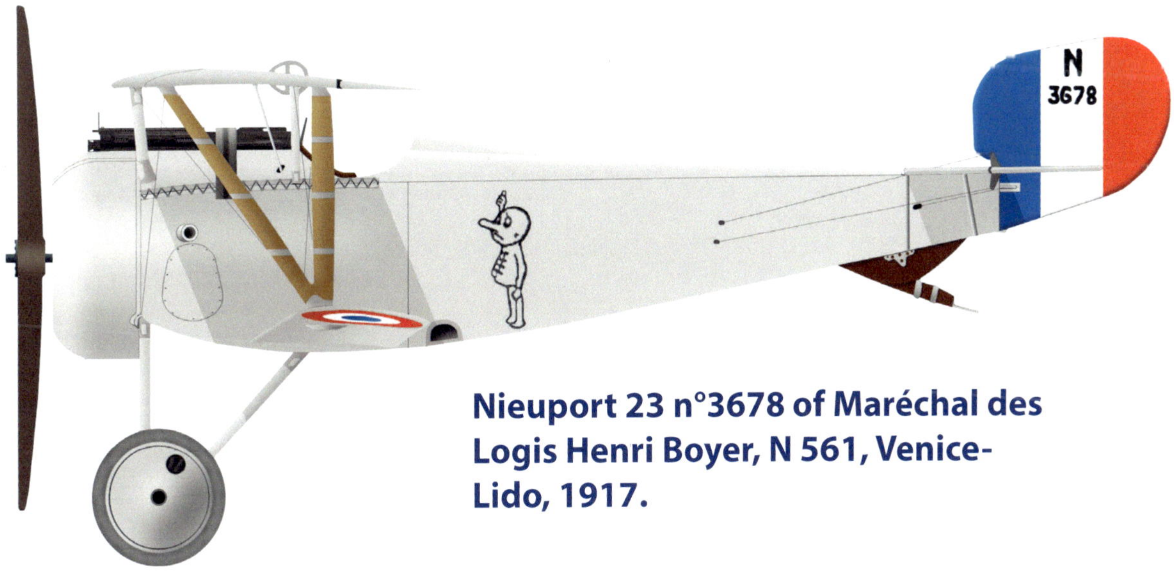

Nieuport 23 n°3678 of Maréchal des Logis Henri Boyer, N 561, Venice-Lido, 1917.

Nieuport 23 (serial unknown) of Sous-Lieutenant Jean Darbos, N 561, Venice-Lido, 1917.

Nieuport 23 (serial unknown) of Sous-Lieutenant Xavier Garros (1 victory), N 561, Venice-Lido, 1917.

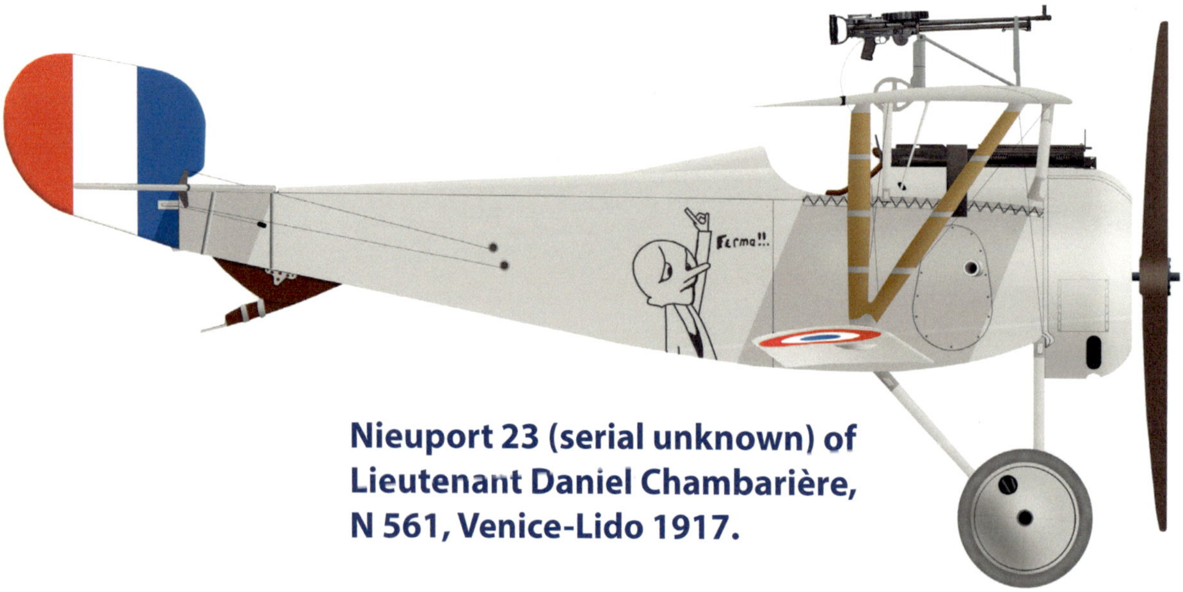

Nieuport 23 (serial unknown) of Lieutenant Daniel Chambarière, N 561, Venice-Lido 1917.

during his missions until his return to France in July 1917. These four pilots of the N 561 were the only ones to approach the enemy during this period, because in Venice, it was so quiet that Captain de Chalonge suggested to his superiors in April 1917 that his squadron be sent back to France or that its numbers be greatly reduced. For diplomatic reasons this was refused, but he was authorized to organize a permanent detachment of four aircraft at Cascina Farello.

There was some excitement in Venice on August 14, 1917, when Austrian naval aviation ventured to carry out a daylight raid. The Nieuports of escadrille N 561 intervened efficiently and won their third confirmed victory by shooting down seaplane K 203 at the mouth of the Piave river. On August 28, a Hansa-Brandenburg fighter was shot down during an escort over the Adriatic. After the departure of the Espinasse squadron, the Venice pilots, who had recovered some Sopwiths 1A2 as well as Lieutenant Olphe-Galliard, a very experienced observer, tried their hand at reconnaissance. A young pilot of the N 561, Maréchal des Logis (cavalry sergeant) Corniglion, served as Olphe-Galliard's pilot on a Sopwith on September 1, 1917, taking off from Cascina Farello for a daring photographic flight over the port of Fiume, 100 kilometers behind the enemy lines. On the way back, their two-seater flew along the west coast of Istria, where it was escorted by the new Macchi M5 fighter planes of the Italian Navy and Nieuport planes of escadrille N 561. At Punta Salvore, the flak shot down a French fighter, Maréchal des Logis Tétard de Morcellier, which was the only loss in combat of the N 561 during the entire war (1 other pilot and an observer were killed in accidents).

On this date, a French logistic base was created in Otranto, in the heel of the Italian boot, where planes in crates were delivered from France by rail. They were then taken off towards the Orient Front and the city of Salonika by a team of convoy pilots who gradually set up a base, studying the safest air routes through the mountains of Greece. A few accidents occurred during these trips but a regular link was effectively set up until the end of the war. The center was attacked by three bombing planes of the Austrian navy on July 25, 1918, which destroyed 12 planes on the ground while they were awaiting delivery, without fortunately causing any victims but showing the importance of this base in the eyes of the enemy.

During the autumn of 1917, 3 other observation squadrons on Sopwith 1A2, SOP 36, 208, and 221, successively passed over the Italian front. Although the French archives are imprecise on their stay, it seems that there were two squadrons of about ten aircraft each at the same time. If we add the 18 aircraft of the N 561, we arrive at a total of 38 combat aircraft, a not insignificant total, but not very important when compared to the 411 aircraft (i.e. 9%) of the Italian air force, of which 12 fighter squadrons were equipped mainly with Nieuport 17s, a few SPAD VIIs intended for elite pilots, and the new Hanriot HD 1, which the Macchi company began to manufacture under license. However, the battle of Caporetto quickly changed the situation.

1.4 - The Battle of Caporetto gives strong reinforcements with French 10th army (November 1917 – April 1918)

On October 24, 1917, a major event took place on the Italian front. Germany came to the aid of its Austrian ally and sent elite troops, planning an offensive that broke through the front near the village of Caporetto (today Kobarid in Slovenia). The Italian army was forced to make a hasty retreat to the west, abandoning a large part of the Veneto and re-establishing a line of resistance on the Piave River, which flows into the Adriatic not far from the Venetian lagoon...

The detachment of escadrille N 561 present at Casina Farello was caught in the middle of the turmoil. It multiplied its missions on the front and even won a victory on October 25 (Aspirant Marcel Robert) against the K 221 seaplane at Miramare. The four Nieuports evacuated to Venice at the last moment, and Lieutenant Loiseau reported in his memoirs that he went to warn a French squadron on a Sopwith by car, which evacuated its airfield shortly before the enemy troops invaded it! The Italian air force in general retreated in a hurry and had to burn many of its unavailable aircraft on the fields left to the enemy, when they were not captured outright. Its strength was halved to 202 operational aircraft according to a count dated November 11, 1917... For the French pilots of the N 561, the front was now very close; they won two confirmed aerial victories against Austrian seaplanes off the coast of Venice on November 16 and 17, 1917, the first one being due to a new recruit, Sergent André Lévy, who won his first certified victory there with his teammate, Maréchal des Logis Edouard Corniglion.

The disaster of Caporetto, which cost General Cardona his head (he was replaced by General Diaz), pushed the French and British governments to send large reinforcements to the Italian front – both to help contain the new front on the Piave and to stimulate the declining fighting spirit of the Italian ally... France sent its 10th army to the Monte Tomba sector, where the Italian troops stabilized the new front. With the ground troops came no less than two fighter squadrons (SPA 69 and N 82) of 15 aircraft each, and 8 reconnaissance squadrons of 10 aircraft each, the AR 14, 22, 44, and 254 (equipped with Dorand AR 1s) as well as the SOP 36, 206, 214, and 221 on Sopwith 1A2s (plus AR 275, formed by splitting AR 44 in February–March 1918). Adding the aircraft of the N 561, now attached to the 10th Army, an air force of more than 130 aircraft supported the 200 or so Italian aircraft that survived Caporetto, reinforced by 3 British fighter squadrons. To bail out the Italian fighter force, which had lost a number of aircraft, nearly 200 Nieuport 27s were delivered to the Italians in two batches of 100 aircraft, allowing the transalpine pilots to hold the front while waiting for the delivery of the Hanriot HD 1s, whose production lines were running at full speed.

A solid logistical system was put in place to receive all these French squadrons, which settled in Verona on November 1917, and then on more advanced bases in San

Above: It was only after the battle of Caporetto, at the end of 1917, when N 561 finally receives some SPAD VII. These models have been in service for more than a year on the French front... The French high command is far from judging the Venice squadron as a priority for its supply of modern aircraft! One of the very first SPADs received, n°283, was appropriated by lieutenant Jean de Limur who decorated it with the Italian version of the Norman motto: *Forse che Si, forse che no* (maybe yes, maybe no). This is in fact the title of a novel published in 1910 by Gabriele d'Annunzio, a regular visitor to the squadron and himself an aviator. (SHD)

Above: A landing by Lt. de Limur on the Lido fort shows the other part of the motto, *"Forse che no"*. (SHD)

Above: Sous-lieutenant Jean Bignon, who won an aerial victory on November 17, 1917, decorated his Nieuport 24bis n°5158 with a magnificent faun playing a flute. In the background is the SPAD VII n°1068 decorated with the Elefante Néro of lieutenant Frédéric Loiseau (SHD)

Above: In the center of the photo, sous-lieutenant Jean Bignon proudly poses in front of a souvenir of his victory against an Austrian seaplane on November 17, 1917 (SHD)

Above: A junkyard of destroyed Italian planes after the retreat of Caporetto in November 1917. In the center we can see a Sopwith 1A2 from escadrille SOP 36. (Paolo Varriale)

Above: Dorand AR n°216 from escadrille AR 14, one of the eight observation units serving with French 10th army sent in Italy to secure the front. (Luigino Caliaro)

Above: Other unit of the French 10th army, escadrille AR 22. Its symbol, difficult to distinguish on the AR 1 aircraft, is the she-wolf suckling Remus and Romulus (Luigino Caliaro)

Below: Autochrome representing the AR 1s of escadrille AR 44, a squadron arriving in Verona on November 5, 1917 with the 10th French Army.

Above: Italian soldier surrounding an AR 2 of escadrille AR 44 that missed its landing. The snowy mountains in the background are clearly visible (Paolo Varriale)

Below: Observation squadron arriving with the French 10th Army, the SOP 214 has a rising sun as its insignia (SHD)

Above & Below: Two photos showing a Sopwith 1A2 from escadrille 221 in Nové, among Italian soliders. (Collection Paolo Varriale)

AR 1 n°366, escadrille AR 14,
Castello di Godego, February 1918.

AR 1 n°205, escadrille AR 22,
Castello di Godego, February 1918.

AR 1 n°215, escadrille AR 44,
Castello di Godego, February 1918.

AR 1 (serial unknown), escadrille AR 254, Verone, February 1918.

Sopwith 1A2 n°1056 (built by Lioré), escadrille SOP 36, Castello di Godego, February 1918.

Sopwith 1A2 n°5173 (built by SACA), escadrille SOP 206, Verone, February 1918.

Sopwith 1A2 n°unknown (built by Hanriot), escadrille SOP 214, Vérone / Porto Nuova, November 1917.

Sopwith 1A2 n°5165 (built by SACA), escadrille SOP 221, Vérone / Porto Nuova, November 1917.

SPAD VII (serial unknown) of lieutenant Honoré de Bonald (5 victories), escadrille SPA 69.

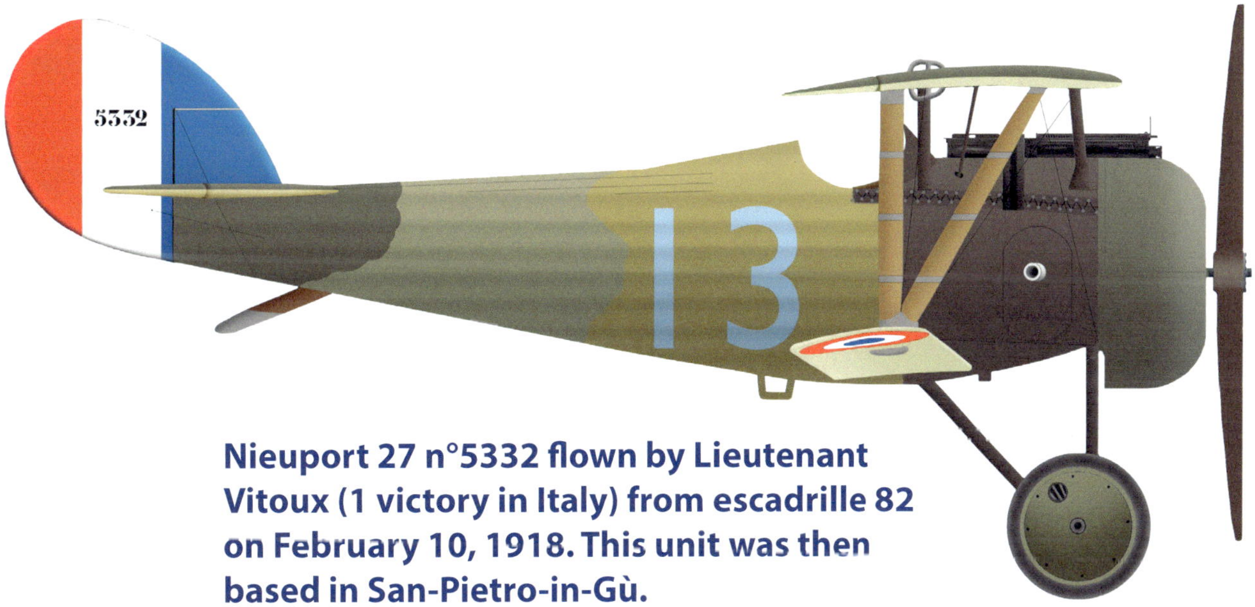

Nieuport 27 n°5332 flown by Lieutenant Vitoux (1 victory in Italy) from escadrille 82 on February 10, 1918. This unit was then based in San-Pietro-in-Gù.

Above & Below:
SPAD VII n°283 of Lieutenant Jean de Limur, escadrille N 561, Lido, Winter 1917–1918.

SPAD VII n°1068 of Lieutenant Frédéric Loiseau, escadrille N 561, Venice-Lido, early 1918.

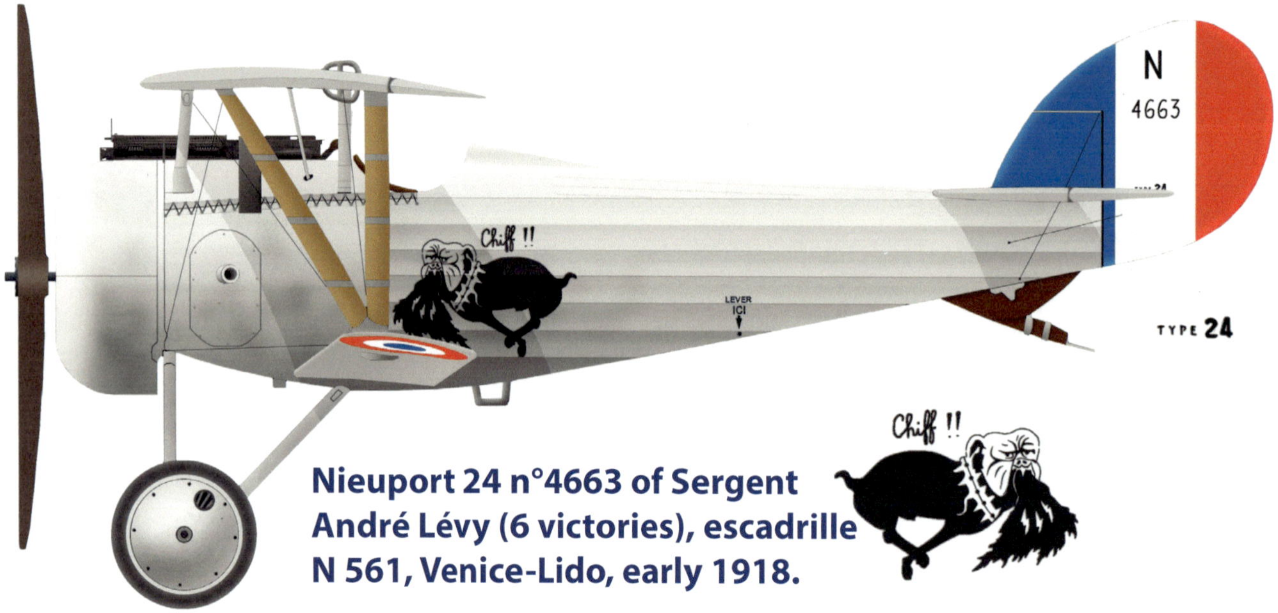

Nieuport 24 n°4663 of Sergent André Lévy (6 victories), escadrille N 561, Venice-Lido, early 1918.

Nieuport 24bis (serial unknown) of Sous-Lieutenant Gabriel Trouvé, escadrille N 561, Venice-Lido, early 1918.

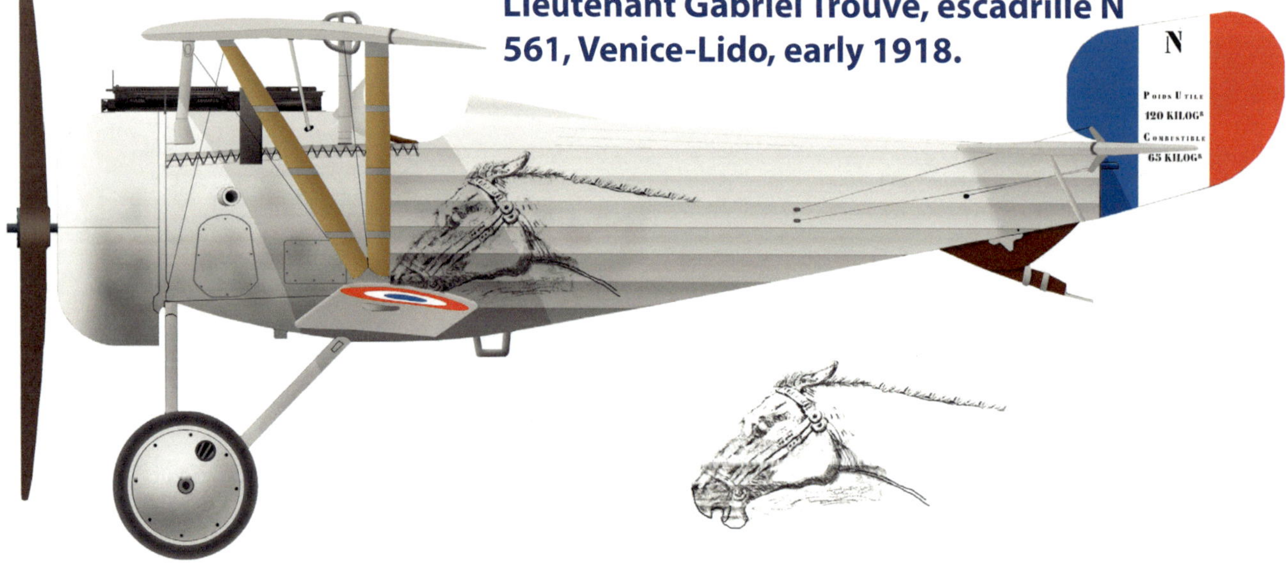

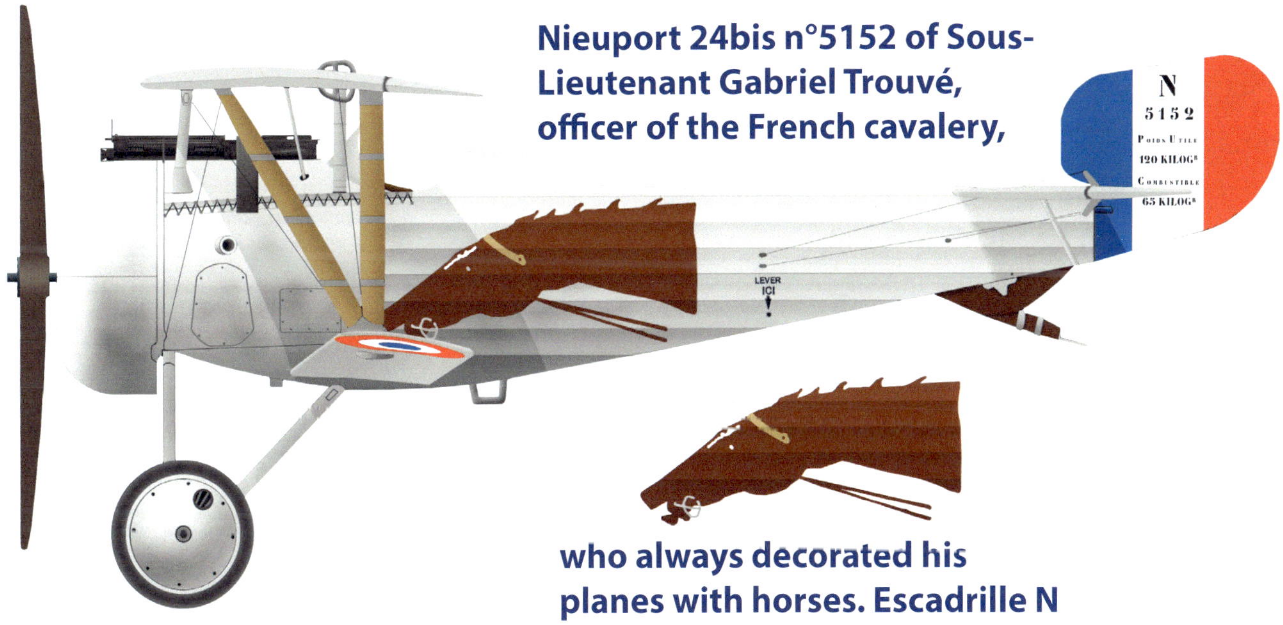

Nieuport 24bis n°5152 of Sous-Lieutenant Gabriel Trouvé, officer of the French cavalery, who always decorated his planes with horses. Escadrille N 561, Venice-Lido, early 1918.

Nieuport 24bis n°5158 of Sous-Lieutenant Jean Bignon (1 victory), escadrille N 561, Venice-Lido, early 1918.

Pietro in Gù (the two fighter units settled there), Castello di Godego, Castelgomberto and Nové di Bassano, which they sometimes shared with Italian squadrons. Their equipment was shipped from France in crates by rail to Milan, reassembled in Brecia and stored in a park located in Verona before being transported to their units. On December 12, 1917, a tragedy struck all French soldiers in Italy when a train carrying soldiers coming back to France derailed in Saint-Jean de Maurienne on a mountain track, killing 425 people, including a mechanic from the Venice squadron, Caporal Georges Duchez - the largest known rail disaster in France to date.

The large mass of airmen of the observation squadrons, flying ARs and Sopwiths, discovered a front very different from the French one, with high mountains rising not far from their fields. This rugged terrain was responsible for several losses, as it made emergency landings impossible in case of engine failure. The cold winter weather did nothing to help, especially for the Sopwiths whose landing gear was very fragile and broke at the slightest inconvenience such as an unseen pebble in the snow... Five airmen lost their lives in accidents during the six-month period spent on the Italian front, but these losses were quite light for a mass of 8 squadrons. On the other hand, a relatively good surprise awaited the French airmen as for the strength of the enemy aviation. This one is seen as not very combative and the air encounters are quite rare compared to France. Only one airman was lost, an observer of AR 14 (Lt Rey) who, wounded in combat, died from his wound on September 30, 1917.

Lieutenant Gabriel de Braquilanges, observer in escadrille AR 254, testified: "On November 1, 1917, we left for Italy where we were going to help the Italians who were getting out of hand after Caporetto. We came to put a stop to it (...) With the Italians, we generally had very good relations, but

Above: Photographed by a Sopwith of SOP 221 which he was escorting, the Nieuport 27 of Captain Raoul Echard, leader of escadrille N 82 and ace with 7 victories (including 3 in Italy), flew over the Piave front on February 25, 1918. Considered obsolete on the French front, the Nieuport 27 nonetheless performed well on the Italian front against an Austrian air force that was not considered very combative. (SHD)

Below: After Caporetto, the Royal Air Force sent 3 fighter squadrons to support the Italians, the 28th, 45th and 66th on Sopwith Camels (on the picture) and later the 139th on Bristol F2B two-seaters, not to mention 2 observation two-seaters on RE 8, the 34th and 42nd Squadrons. Very aggressive, the British pilots were characterized by their very lax system of homologation of victories, which made them claim more victories for their 3 fighter squadrons than the entire Italian army did during the entire war! (Paolo Varriale)

we did not consider them to be very bellicose, especially at the beginning. We didn't deal with that in the air force, but I heard soldiers of the 126th RI that I often went to see tell me that they had been obliged to set up guard posts along the roads to stop the Italian soldiers who were getting away! And they were Italian officers who went to the rear by car or on horseback: la guerra e finita! We were obliged to stop all these people, otherwise they would go back to Sicily...

I went to escadrille 254 as a photographer. In addition to the artillery adjustments, I was in charge of making photographs of some enemy locations. I flew with several pilots, but I remember one of them, particularly remarkable, S/Lt Lucien Saflix, who saved our asses by landing on a river bed of the Po plain after an engine failure – on a strip of sand of 5 or 6 meters! (...) In mission we had to fly as high as possible and we could not go beyond 4000 m, where because of the rarefaction of the air we were sweating profusely at the slightest effort, including the handling of the camera (...)

The Austrians who were in front of us were not very strong (...) They did not come out much, I do not know if they were less numerous, in any case we did not see them much. But they had remarkable artillery, especially heavy artillery. Every time I went on a reconnaissance mission to an important target, I came back with holes in my canopy. This pleased the mechanics, who would plug them up with pellets and say that their pilot had fought well! I remember taking a picture of Mount Asiago, far away in the lines, from where I came back with a lot of shards in the canopy of the plane.

As there were several squadrons on the same field, on rainy days when we were not flying, we played poker with the Italian pilots at the table. The civilians liked us a lot. At the funeral of a French pilot whose engine burned out in flight, a large crowd of Italian soldiers and civilians came. As French airmen, we were particularly well regarded. It must be said that when we were on the Verona airfield, the inhabitants could only see us, as our airfield was close to the city, while the infantrymen were camped elsewhere.

We had connections with the Italian airmen, as well as with the British who had done the same after Caporetto and sent troops. The English airmen only had fighters. We realized that they did things well. Once we saw an English fighter coming in zigzagging, and landing in a hurry in the opposite direction of the road. We ran over with the ambulance: the guy was almost dead and had managed to land! I think he made it..."

The same was true of the fighter pilots of the two squadrons sent with the 10th Army, SPA 69 and N 82: the difficulties encountered did not come from the enemy air force, which commandant Houdemon, appointed head of the French air force in Italy, judged as "not very biting". The losses suffered, which were very slight, were mainly due to the winter cold and the terrain, which was "very unsuitable for aviation". The Austrian air force injured only two pilots, while the accidents killed one and injured nine others... Escadrille SPA 69, whose symbol is a cat's head, was equipped with the powerful SPAD VII and received in March 1918 some SPAD XIII which were the best fighter aircraft of the time. The men of Captain Paul Malaviale, an ace with 5 victories, shot down 4 Austrian planes during their Italian stay and did not hesitate to directly search for the enemy on its grounds by strafing the enemy hangars. As for escadrille N 82, it seems to have been neglected by the quartermaster's office because it was equipped with Nieuport 24 bis and Nieuport 27 which were considered to be outdated on the French front. Houdemon notes about them: "They were what they are in France: regular aircraft, inferior to the enemy Albatros fighter." Their obsolescence did not prevent them from performing well on the Italian front against the Austrian air force, since they won 7 air victories, 3 of which were obtained by Count François de Boigne, a lieutenant-pilot who ended the war with 6 victories.

1.5 - A limited presence during the battle of the Piave and the final offensive

The German spring offensive launched in France in April 1918 led the 10th French Army to hastily pack up the Italian front with most of its squadrons. The 12th Army Corps remained on guard on the Piave, with its two organic squadrons, the AR 22 and AR 254, which received Salmson 2A2s in May 1918. Based at Nové, they benefited from the fighter escort of 4 aircraft of the N 561, which maintained a permanent detachment there.

The French air force, whose strength now amounted to 38 aircraft, was only a very modest auxiliary force for the Italian air force, which in June 1918 had completely healed its wounds: the Corpo Aéronautico Militare fielded nearly 550 aircraft in the front line, including 221 fighters in 13 squadrons, many of which flew the Hanriot HD 1, which was much appreciated by the pilots for its maneuverability.

In addition to these land-based aircraft fighting on the front line, the seaplanes of the maritime air force were also present in the Sicilian Straits. In January 1918, the Marsala Battle Station was transformed into an autonomous Maritime Aviation Center under the command of Ensign Clarice. Its strength increased to 12 armed aircraft, some Donnet-Dennaut 200 hp Hispano, some Tellier-Canon 200 hp Hispano and, towards the end of the summer 1918, some Georges Lévy 280 hp Renault. On January 27, 1918, one of its patrols attacked a suspicious wake near an Italian sailing ship damaged by a U-boat. Bombs in the water like most of the seaplane attacks of WWI... On 8 July 1918, another U-Boat was spotted on the surface and attacked. Three bombs fell near its periscope but the U-Boat, probably shaken, managed to escape by diving. At the same time as the activity of the CAM Marsala, a four-aircraft combat post was opened on the island of Lampedusa, depending on the CAM Sousse in Tunisia. Its Donnet-Dennaut 200 hp Hispano seaplanes took part in the escort of convoys coming from Malta.

Above: Presumably in the spring of 1918, maréchal des logis André Loyseau de Grandmaison decorated his Nieuport 24 n°4393 with a splendid dragon. He was already wearing such a decoration on his previous Nieuport 23. This pilot won a victory on August 28, 1917 by shooting down an Austrian Hansa-Brandenburg CC fighter plane in the Adriatic Sea (SHD)

When on June 15, 1918, the Austro-Hungarian army launched its last chance offensive by attempting to cross the Piave River, the Italian fighters, employed on the battlefield (Massa da Caccia provisional with 8 Squadriglie) under the command of Pier Ruggiero Piccio promoted to the rank of tenente colonello, had no difficulty in gaining control of the air in the face of an enemy air force handicapped by all sorts of shortages, and quickly swept out of the sky, allowing Italian bombers to attack Austrian pontoons on the Piave River and reconnaissance planes to direct the artillery fire. A storm of fire fell on the Austro-Hungarian soldiers who returned to their initial positions with heavy losses. The

Below: Nieuport 27 in aluminium livery, received by the Venice squadron at the beginning of 1918. The unidentified pilot who owned it had it decorated with a beautiful panther (Paolo Varriale)

Above & Below: Presumably Maréchal des logis Edouard Corniglion had his last Nieuport 27 (n°5239) decorated with his favorite imprecation, "Ocio - Fiol d'un Can!", using both sides of the aircraft's fuselage, which can be seen on patrol over the Venetian Lagoon. Sick of malaria, which was endemic in the Italian lagoons at the time, Corniglion left the squadron in June 1918, having won two certified victories. He would win four more in 1940 at the controls of his Curtiss H-75, then second in command of Fighter Group III/2! Film producer, journalist, raid pilot and personal friend of André Malraux during the interwar period, he joined the resistance after the 1940 campaign and joined the Free French Air Force, fighting again in North Africa. Finishing with the rank of general, he became congressman and minister of transport and then of justice in the governments of the French Fourth Republic. This companion of the liberation died in 1963 (Paolo Varriale)

Above: One of the best pilots of the N 561 squadron with two certified victories, sous-lieutenant Xavier Garros poses in the Nieuport 27 received by his unit in early 1918. He had it decorated with a personal insignia, a Pinocchio head (SHD)

Below: During the summer of 1918, an inspection took place at Nové where the Salmson 2A2s of escadrille SAL 22 were located. In the center of the picture, in light pants, the young Prince of Wales (future Edward VIII, King of England) walks. Immediately to the right, the King of Italy Victor-Emmanuel III, then General Armando Diaz, Chief of Staff of the Italian Army. (Lugino Caliaro)

Above: The young Prince of Wales was received at the Villa Paradiso and talked with the pilots, under the amused eye of capitaine de Chalonge, cigarette in hand. The pilot Frédéric Loiseau wrote about this in his memoirs: "The Prince of Wales, arriving one day for lunch, won us all over with his royal simplicity. Commander Accorretti had been asked to accompany the Prince as he passed by. The day before, he came to see me. It was a question of anticipating the case where H.R.H. would wish to pay one of his visits that Robert de Flers said were "reserved for the president of the senate", for crowned heads passing through Paris [Author's note : a term used by protocol to allow a head of state to discreetly visit a reputed brothel in Paris such as the "One two two" or the "Chabanais"]. He was advised by his superiors : "If you are embarrassed, contact the French squadron! "Our reputation, as you can see, was well established. We called on several "authorized" advisers. We chose a "little ally" who, in addition to a good physical reputation, had the advantage of speaking French and even a few words of English. It is not for me to say whether the Prince did follow through…" (SHD)

Right: Sergent André Lévy's favorite target was the observation balloons that rose above the Austrian lines in the Piave plain, an indispensable tool for observing enemy positions. He shot down five of them during the summer of 1918. If the balloon represented on the photo is in fact Italian, the landscape it flies over is indeed the marshy plain of the Piave. (Paolo Varriale)

Above: After his 5th balloon destroyed, sergent Lévy was not lucky, as his plane was hit by flak and had to land in the enemy lines where he deliberately wrecked his plane. The photo, taken by Austrian soldiers, shows the personal decoration of his SPAD XIII n°4789: a bulldog head, drawn in memory of a dog he owned and liked. (Paolo Varriale)

Facing Page, Below: On November 6, 1918, Sergent André Lévy (standing in the 1st row, 4th from the right) returned to his squadron. The ace of Venice with 6 victories escaped from Austria by crossing snowy mountains on foot... He became a reserve officer during the interwar period and served as a cipher officer during the 39-40 campaign. He died on March 12, 1973 (SHD)

Below: View of the aircraft of N 561 lined up on the Lido airfield, in January 1919, shortly before the squadron was disbanded. In the foreground, the two two-seater SPAD XVIIs of the squadron intended for long-range reconnaissance missions. An examination of the fighters in the background shows that the unit was equipped with half SPAD VIIs and XIIIs, and half Nieuport 27s, aircraft that were considered obsolete and relegated to second lines units. (SHD)

Above: André Lévy after the armistice.

Left: André Lévy back from captivity, November 1918.

Nieuport 24 n°4393 of maréchal des logis André Loyseau de Grandmaison (1 victory), escadrille N 561, Venice-Lido, 1918.

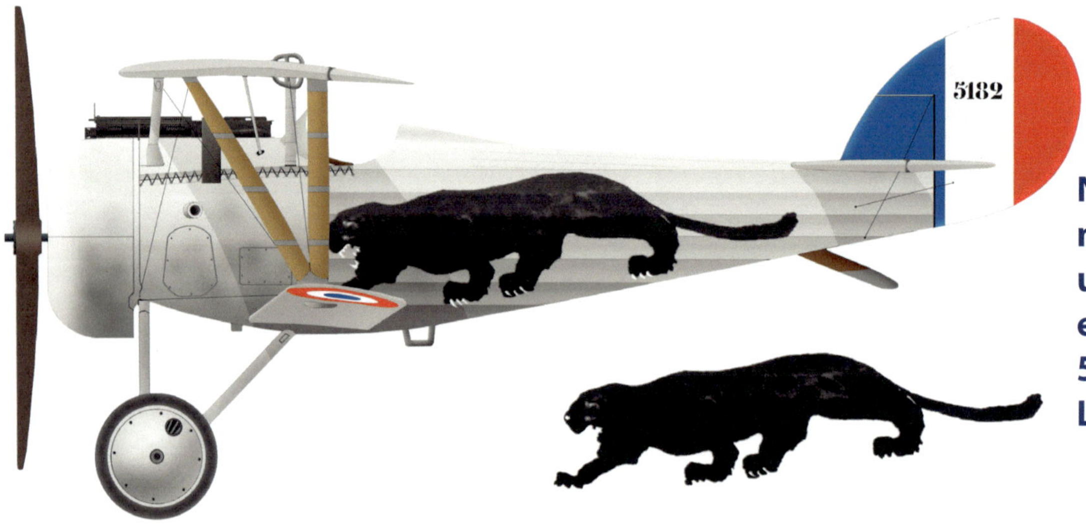

Nieuport 27 n°5182, pilot unknown, escadrille N 561, Venice-Lido, 1918.

Nieuport 27 n°5239 of maréchal des logis Edouard Corniglion (2 victories), escadrille N 561, Venice-Lido, 1918.

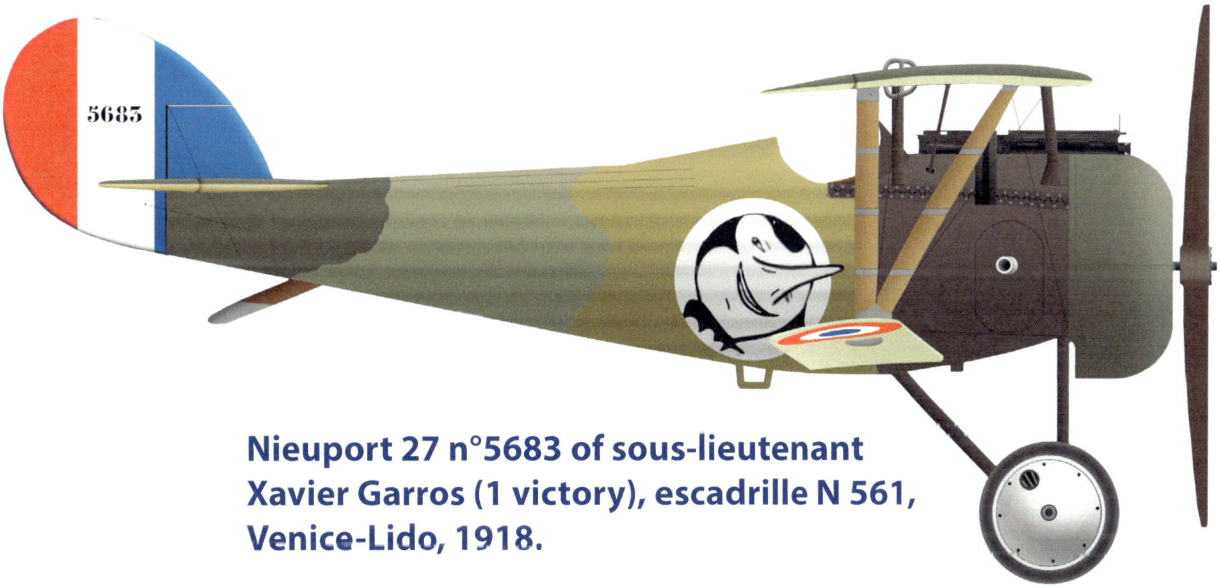

Nieuport 27 n°5683 of sous-lieutenant Xavier Garros (1 victory), escadrille N 561, Venice-Lido, 1918.

SPAD XIII n°4764 of maréchal des logis André Lévy (6 victories), escadrille N 561, Venice-Lido, October 1918.

SPAD XI n°6123 of sergent Séraphin Rapello, escadrille N 561, Venice-Lido, 1918. N 561 always kept a section of two-seaters for strategic reconnaissance missions.

Salmson 2A2 (serial unknown), escadrille SAL 22, Nove-di-Bassano, Summer 1918.

French air force only participated symbolically in the battle, when on June 22, 1918, three French pilots, led by Sergeant André Lévy, a pilot as tough as the bulldog adorning his aircraft, managed to destroy an enemy observation balloon with the help of two Italian fighter planes.

After their failure on the Piave, the Austrians did not attempt any further large-scale action. But in Venice, André Lévy developed a taste for the dangerous sport of Drachen hunting, and from then on he patrolled the Piave River to find others... He succeeded in destroying two of them on July 20, then another one in collaboration with a teammate on August 5, thus becoming the "ace of Venice". The Drachen he destroyed on 16 September did not bring him luck, as his SPAD XIII was hit by flak and he had to land in the nearby Austrian lines, where he was captured. Interned in a prison camp, he quickly deserted his hosts and after two escape attempts managed to reach the Italian lines on foot through the snow-covered mountains, and rejoined his squadron on November 6 where he was celebrated as a hero.

On that date, the old Austro-Hungarian Empire ceased to exist. Undermined by its internal nationalist tensions, its troops collapsed during the final assault launched by Italian troops on October 24, 1918 and known as the battle of Vittorio Veneto. A Massa da Caccia took over and quickly overran the last Austrian fighters, who succumbed to the numbers. The Caproni bombers were given free rein to intervene in the key sectors of the attack and help to achieve the decisive breakthrough. The 12th French Corps took part in the attack and its two observation squadrons deplored the loss of one observer, Adjudant Léon Bigot, killed by a shot from the ground at Vittoréo. Lieutenant Gabriel de Braquilanges, still stationed at SAL 254, recounted: "At the battle of Vittorio Veneto, the day before I had just bombed an Austrian airfield. We had 75 mm shells attached to our wings (...) Three days later the armistice came. As I had seen my bombing impacts, I took a car (we had two in the squadron) and I left to see the airfield I had bombed. I was able to get there despite the roads being in a deplorable state and I saw the hangar I had targeted, which was hit. There was a demolished plane in it, I thought it was mine and cut the fuselage canvas as a souvenir.

At the end of the war, our corps attacked at Asiago. There, the guys from the 126th Infantry Regiment told me that the Italians had recovered well from Caporetto and fought well. The aerial fighting was less violent than on the French front, much rarer because the Austrians did not go out much. Even on the ground, there was no comparison with the Boches. The Austrians were made up of all sorts of nationalities. After the armistice, I returned to the infantry and took command of a battalion. We left Venice for Cattaro [on the Yugoslav coast], where I finished the war. There, I had to make a protocol visit to the commander of the local base. I had to deal with one of my former enemies: he was a Serb who had been mobilized as an Austro-Hungarian. He told me that he had been in front of me in Asiago, I had helped to fire on his batteries! Many of these guys were only thinking about their independence, which they got later."

The armistice on the Italian front came into effect on November 4, 1918, one week before Germany's surrender in France. During these four years of combat, the French air force played a sometimes modest, but sometimes important role at key moments in the fighting, achieving a total of 24 certified victories for only 7 killed in action and 2 prisoners. The help to the Italian ally was mainly manifested from a technical point of view by the sale of aircraft or production licenses which allowed the Italian air force to be equipped with a large majority of French aircraft, including all of its fighter aircraft. However, little would remain of this brotherhood of arms due to nationalist tensions and economic rivalry between the two nations, which became enemies during the Second World War. Maréchal des Logis Edouard Corniglion, who became General Corniglion-Molinier, commander of the Free French Air Force, faced his former comrades in arms at the controls of his Bristol Blenheim over Tobruk in 1941...

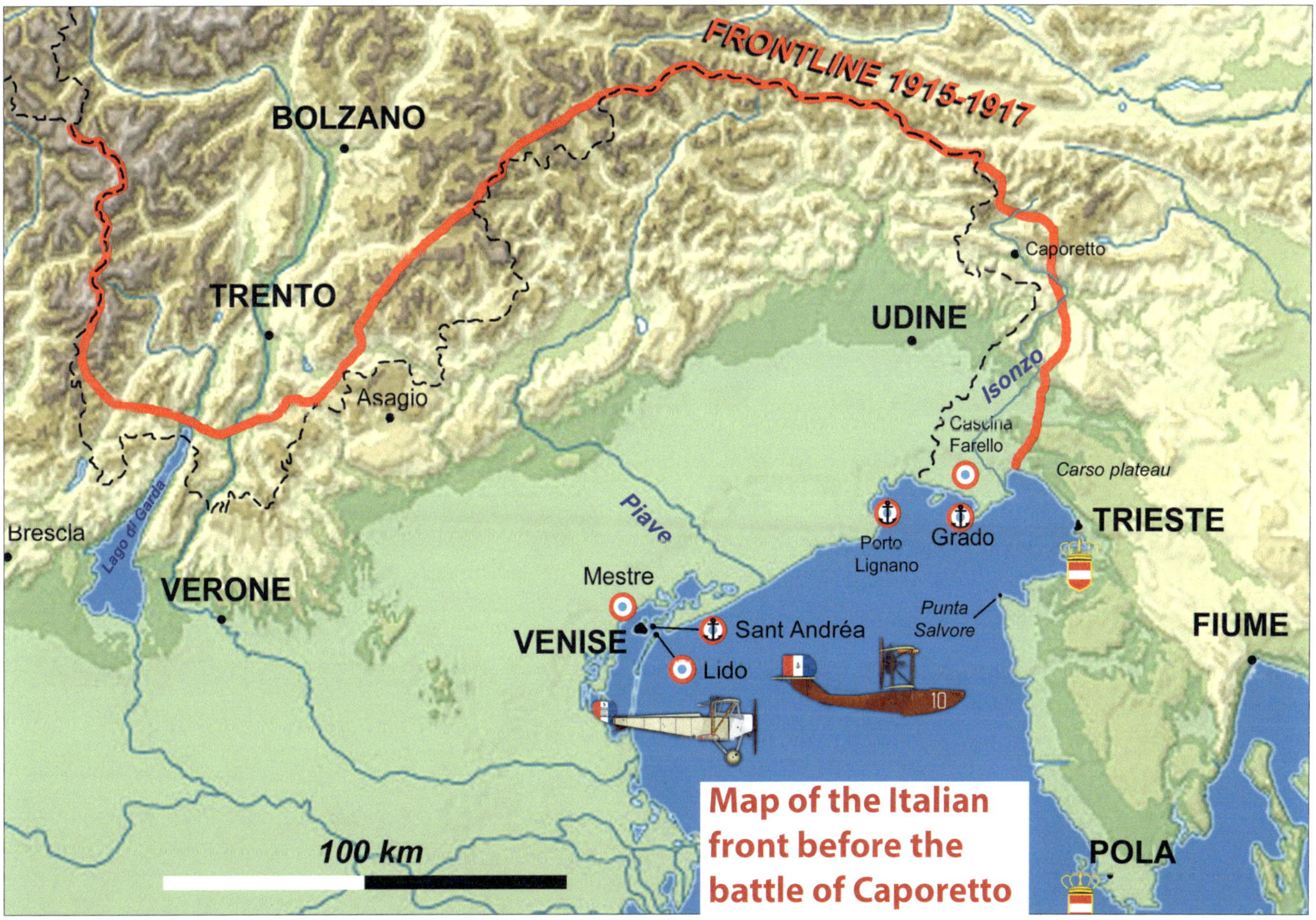

Map of the Italian front before the battle of Caporetto

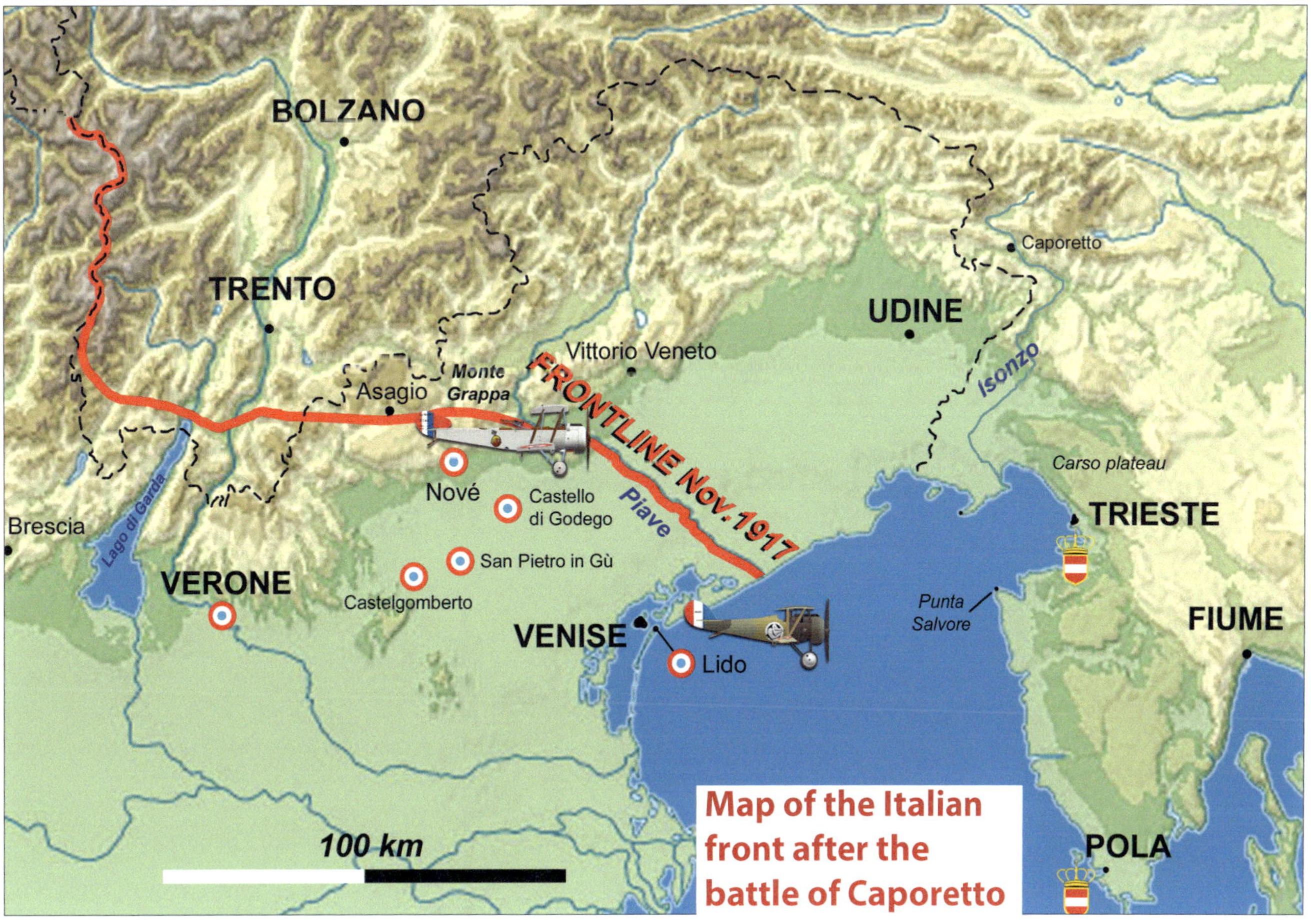

Map of the Italian front after the battle of Caporetto

Victory Claims on Italian Front by French Pilots 1915–1918

#	Date	Rank	Pilot or Gunner		Unit	Plane Piloted	Plane Claimed
-	9/13/16	LT	Chambarière, Daniel	-	N 561 - N 392	Nieuport	Seaplane
1	10/23/16	S/Lt	Garros, Xavier	1	N 561 - N 392	Nieuport	Lohner seaplane L 138
		STV 2e Capo	Martinengo, Carlo (pilot) Mincotti, Daniele (gunner)			Macchi L 2	
-	10/23/16	Brig	Corniglion, Edouard	-	N 561 - N 392	Nieuport	Lohner seaplane
-	4/17/17	S/L	Garros, Xavier		N 561 - N 392	Nieuport	Seaplane
		Lt Sol	Robert, René (pilot) Salbreux, François (Gunner)		N 561 - N 392 N 561 - N 392	Nieuport 12	
-	5/13/17	MdL	Corniglion, Edouard		N 561 - N 392	Nieuport	Seaplane
	6/21/17	MdL	Martin, Pierre (pilot)		Espinasse	Sopwith 1A2	Airplane
		S/Lt	Olphe Galliard, Louis (Observer)		Espinasse		
	8/14/17	MdL	Geoffre de Chabrignac (-de), Pierre		N 561	Nieuport	Seaplane
2	8/28/17	MdL	Loyseau de Grandmaison, André		N 561	Nieuport	Seaplane
	9/4/17	MdL	Geoffre de Chabrignac (-de), Pierre		N 561	Nieuport	Seaplane
	9/4/17	MdL	Geoffre de Chabrignac (-de), Pierre		N 561	Nieuport	Seaplane
3	10/25/17	S/Lt	Robert, Marcel		N 561	Nieuport	Seaplane
4	11/16/17	Sgt	Lévy, André		N 561	Nieuport	Seaplane
		MdL	Corniglion, Edouard		N 561	Nieuport	
5	11/17/17	S/Lt	Bignon, Jean		N 561	Nieuport	
		Guardiamarina	Morterra, Paolo		Sq 260	Macchi M 5	
		2e Capo Timoniere	Zanetti, Arturo		Sq 260	Macchi M 5	
	11/20/17	MdL	Geoffre de Chabrignac (-de), Pierre		N 561	Nieuport	Seaplane
6	**1/1/1918**	Sgt	Balzac, Billy		SPA 69	SPAD	Two-seater
	1/8/18		Unknown crew		?		Two-seater
-	1/10/18		Unknown crew	-	AR 14	Dorand AR	Fighter

Place	Source	Confirmation	Notes
Adriatic sea, between Venice and Parenzo	French report	Considered as forced to land	Seaplane K 163 damaged - then destroyed on the ground during repairs
Laguna Caorle (pente Tagliamente)	Italian report	Citation OA 15-11-1916 - JORF 09-03-1917 p.1890	Lohner L 138 captured - Pilot seekadet Frantz Viranyi POW and observer Freiw. Hoch POW of KIA
Laguna Caorle (pente Tagliamente)	French report		Observer seen wounded
15 Km East Venice		No confirmation by French authorities	Seaplane K 192 forced to land (pilot Planchner, Obs Szechny), towed by Austrian destroyer
Chioggia			
?		Italian decoration, but no confirmation by French authorities	
Adriatic sea		No confirmation by French authorities	Seaplane K 203 forced to alight. Crew saved by austiran boats, seaplane destroyed
Golfe de Trieste		Citation OA 30-08-1917 - JORF 18-12-1917 p.10354	Hansa-Brandenburg CC "A 14" - Seefähnrich Bruno Von Stransky-Stanograd (KIA)
Punta Salvore		Considered as forced to land	
Punta Salvore		Considered as forced to land	
Miramar (near Pelinte Salvatore)		Citation OA 01-01-1918 - JORF 31-01-1918 p.1109	Seaplane K 221 - Pilot Ottinger, Obs Gerszetic (both KIAs)
Cortellazo		Citation OA 12-12-1917 - JORF 22-01-1918 p.812	Seaplane K 374. Pilote Röver KIA (head shot), copilot UNH
		Citation OA 12-12-1917 - JORF 22 01 1918 p.812	Seaplanes K 371 and K 373 forced to land, K 371 sinks
Mouth of Piave river		Citation OA 12-12-1917 - JORF 22-01-1918 p.811-812	Seaplane K 211. Pilot Fregattenleutnant Julius Fedrigoni Edl. Von Etschal (POW), Obs Preiss (POW)
			Fedrigoni captured by Morterra who took him on his M 5.
			Preiss captured by a Macchi L 3. Two other Austrian seaplane flee.
		Considered as forced to land	Seaplane K 161 (Pilot Kaudek, Obs Lupinski) MIA
Alano-di-Piave region	French 10th army report	Citation OA (MM) 01-04-1918 - JORF 23-08-1918	14:00 / 14:10. Claimed as probable
Asiago region	French 10th army report	No confirmation by French authorities	Seen diving, observer wounded
Velstagna region	French 10th army report	No confirmation by French authorities	Seen diving out of control

Victory Claims on Italian Front by French Pilots 1915–1918

#	Date	Rank	Pilot or Gunner		Unit	Plane Piloted	Plane Claimed
7	1/28/18	Adj	Rimbaud, Jean Camille	2	N 82	Nieuport 24 or 27	Airplane
			Italian pilot				
8	2/24/18	Cne S/Lt	Echard, Raoul	3	N 82	N. 24 or 27	Airplane
			Boigne (-de), François	2	N 82	N. 24 or 27	
9		Cne	Echard, Raoul	4	N 82	N. 24 or 27	Airplane
		S/Lt	Boigne (-de), François	3	N 82	N. 24 or 27	
-	2/25/18		Unknown Pilot		N 82	N. 24 or 27	Airplane
10	3/13/18	MdL	Glarélial (-de), Marcel	1	N 82	N. 24 or 27	Fighter
11	3/16/18	Lt	Bonald (-de), Honoré	5	SPA 69	SPAD	Airplane
		Sgt	Balzac, Billy	-	SPA 69	SPAD	
12	3/18/18	Adj	Delelée-Deshayes, Joseph	1	N 82	N. 24 or 27	Two-seater
13		LT	Vitoux, Pierre	1	N 82	N. 24 or 27	Two-seater
		Cne	Echard, Raoul	5	N 82	N. 24 or 27	
14	3/19/18	Cne	Malaviale, Paul		SPA 69	SPAD	Fighter
15	3/21/18	Adj	Franc, Joseph	1	SPA 69	SPAD	Fighter
-	3/23/18	S/Lt	Boigne (-de), François		N 82	N. 24 or 27	Airplane
-		Lt	Flers, Robert		N 82	N. 24 or 27	
16	3/25/18	S/Lt Sgt	Boigne (-de), François Boland, André	4 1	N 82 N 82	N. 24 or 27	Two-seater
17	6/21/18	Sgt MdL MdL S. Ten S. Ten	Lévy, André Corniglion, Edouard Boyer, Henri Ravazzoni, Ivo Pagliacci, Guiseppe	2 2 1	N 561 N 561 N 561 Sq 260 Sq 260	SPAD SPAD SPAD Macchi M 5 Macchi M 5	Drachen
-	7/15/18	S/Lt	Darbos, Jean	-	N 561	SPAD	Airplane
18	7/20/18	Sgt	Lévy, André	3	N 561	SPAD	Drachen
19	7/2018	Sgt	Lévy, André	4	N 561	SPAD	Drachen
20	1/8/18	Sgt MdL	Lévy, André Autissier, Charles	5 1	N 561 N 561	SPAD SPAD	Drachen
21	9/16/18	Sgt	Lévy, André	6	N 561	SPAD	Drachen

Place	Source	Confirmation	Notes
	French 10th army report	Citation OA 28-03-1918 - JORF 23-05-1918 p.4510	
Grappa / Monte Pallone - sector Monte Tomba	French 10th army report	Citation OA 28-03-1918 - JORF 23-05-1918 p.4510 "	
Grappa / N. Possagno - Sector Monte Tomba	French 10th army report	Citation OA 28-03-1918 - JORF 23-05-1918 p.4510	
		"	
S. Montello	French 10th army report	No confirmation by French authorities	
Roana-Rodzo	French 10th army report	Citation OB 16-04-1917	11h45
	French 10th army report	Citation OA 28-03-1918 - JORF 23-05-1918 p.4511	Rapport SPA 69 + Rapport Xe armée
NE Roana	French 10th army report	Citation OA 21-05-1918 - JORF 01-07-1918 p.5717	9:00 - Shot until it reached the ground
N. road Rotzo-Roana	French 10th		
	army report	Citation OA 10-10-1918 - JORF 26-05-1919 p.5423 "5th victory"	
Sector Belfioro / W. Cesotto		Citation OA 20-09-1918 - JORF 18-05-1919 p.5102 "5 victories"	
N. Gallio	French 10th army report	Citation OA 01-01-1918 - JORF 31-01-1918 p.1109	11h00
Montello region	French 10th army report	No confirmation by French authorities	11h45
NW. Asagio	French 10th army report	No confirmation by French authorities	11h55. Seen diving in the lines.
N. Montebelluna	French 10th army report	Citation OA 21-05-1918 - JORF 01-07-1918 p.5717 "4th victory"	Wreck found in Italian lines
		Citation OA(MM) 01-08-1918 - JORF 11-08-1918 p.7082 Citation OA(MM) 10-07-1918 - JORF 26-07-1918 p.6454	Ballonkompagnie 20. Observer : Ltn Alexander Pfau (OK). Drachen riddled with 124 bullets.
		No confirmation by French authorities	
		Citation OA 20-08-1918	Seaplane K 161 (Pilot Kaudek, Obs Lupinski) MIA
Alano-di-Piave region		Citation OA 20-08-1918	
Noventa (near Piave river)		No confirmation by French authorities	Ballonkompagnie 11. Destroyed in flames at 9h50 according to austrian archives
W. Ceggia			Ballonkompagnie 3, crew Frantz Nedbalek and Fw Albert Markofsky (UNH). Lévy captured.

French Losses on the Italian Front 1915–1918

Date	Pilot (P) or Observer (O), Fate	Unit	Plane Piloted
6/26/16	EV 1 Vaugeois André (POW) + Observer Grammaticopoulo (Italian) (KIA)	CAM Venise	FBA C n°94
15-Aug-16	SM Duclos Jules (OK) + QL Le Men (OK)	CAM Venise	FBA H n°310
15-Aug-16	EV 1 Roulier Jean (KIA) + QM Costerousse Auguste (KIA)	CAM Venise	FBA H n°308
18-Oct-16	Adj Melin André (WIAcc)	N 561 - N 392	Nieuport 16 n°417
16-Nov-16	Lt Virolet Marcel (WIAcc)	N 561 - N 392	Nieuport n°1688
17-Apr-17	EV 1 Woltz André (MIA) + QM Coste Maurice (MIA)	CAM Venise	FBA H n°384
4-Sep-17	MdL Tétard de Morcellier Jean (KIA)	N 561	Nieuport
15-Dec-17	Cne Lavergne (WIAcc)	Chef aviation 31e CA	
27-Dec-17	MdL Faure Gérard (WIAcc)	N 82	Nieuport 24 or 27
30-Dec-17	Lt Bonnot (OK) + Lt Rey (WIA-DOW)	AR 14	Dorand AR 1
11-Jan-18	Sgt Lenis (KIAcc) + S/Lt Chouk (WIAcc)	AR 214	Dorand AR 1
27-Jan-18	MdL Lemaitre Jacques René (KIAcc) + Lt Douffiagnes Georges (KIAcc)	SOP 214	Sopwith 1A2
27-Jan-18	Adj Chalambel Jean (WIAcc)	N 82	Nieuport 24 or 27
2-Feb-18	Lt Mathis (WIAcc)	SPA 69	SPAD
10-Feb-18	Adj Welling-Nielsen Hans (KIAcc) + S/Lt Cotrait (WIAcc)	SOP 221	Sopwith 1A2
13-Feb-18	Adj Vaillant André (WIAcc)	N 82	Nieuport 24 or 27
14-Feb-18	Lt Wambergues Jean (KIAcc)	SOP 214	Sopwith 1A2
24-Feb-18	Brig Champion Jacques (OK)	N 82	Nieuport 24 or 27
25-Feb-18	Lt Poivilliers Joseph (KIAcc)	AR 22	Dorand AR 1
26-Feb-18	Sgt Soulème Gustave (KIAcc)	N 82	Nieuport 24 or 27
10-Mar-18	Brig Champion Jacques (WIAcc)	N 82	Nieuport 24 or 27
13-Mar-18	MdL Berthelot (WIA) + Lt Ladevèze (WIA)	AR 44	Dorand AR 1
20-Mar-18	Cprl Cottaz Francis (WIAcc)	SPA 69	SPAD
17-Apr-18	MdL Nacrier (KIAcc) + Lt Poitevin (WIAcc)	AR 254	Dorand AR 1 n°522
23-Apr-18	Sgt Marcou Yvan (WIA)	N 561	SPAD
8-Aug-18	Cprl fourrier Regnac Henri (WIAcc) + 2e Cl Derain Louis (WIAcc)	N 561	Sopwith 1A2
16-Sep-18	Sgt Lévy (POW)	N 561	SPAD XIII n°4764
25-Sep-18	Sgt Rapello Seraphin (WIAcc)	N 561	SPAD XI
12-Oct-18	Mdl Autissier Charles (KIAcc)	N 561	SPAD
30-Oct-18	Pilot unknown (OK) + Adj Bigot Léon Louis (KIA)	AR 22	Dorand AR 1

Note : On October 31, 1916, a bomb falls from the wing of FBA H n°321 a kills 2 pilots of CAM Venise, LV Antoine Reynaud

Cause	Place	Notes
Aerial combat	Gulf of Trieste	Banfield
Aerial combat	Gulf of Trieste	Banfield - Plane and crew towed back to italian lines.
Aerial combat	Gulf of Trieste	Banfield - Wreck captured by austrian navy.
Engine malfuction - crash	Laguna of Venice	16h00.
Engine malfuction - crash	Lido Airfield	16h00.
Flak	Adriatic sea	Shot down by flak from austrian ship.
Flak	Punta Salvore	
Crash on landing		
Engine malfunction - crash	Bassano	
Aerial combat		DOW 31-12-1917
Accidental spin	Castelgomberto	
Takes fire on flight, crashes on landing	Trissino	
Engine fire on takeoff	San Pietro in Gu	
Engine malfunction - crash	Zugliano	
Crash on takeoff	Nové	
		No combat loss, according to 10th army document
	Castelgomberto	
Crash on takeoff	San Pietro in Gu	
Accidental bomb explosion	Castello di Godego	
Engine malfunction - crash	Citadella	
Flak	Pasubio	
Crashes on landing	Vicence	
Engin takes fire on takeoff	Nové	
Flak	San Dona Di Piave / Motta di Livenza	9h30. Wounded on his feet
Bad landing	Lido Airfield	18h30. Lands on the roof of a barrack
Flak	W. Ceggia	Escaped shortly before the end of the war
Crash on landing	Lido Airfield	10h10.
Air collision with Macchi M 5	Laguna of Venice	
Flak	Vittoréo	

iat, plus 13 french mechanics, 3 italian mechanics, and wounds 8 french plus 9 italian mechanics.

Chapter 2: French Flyers in Russia

ИМПЕРАТОРСКІЙ

Всероссійскій Аэро-Клубъ

Above: The logo of the IVAK, (Imperatorskii Vserossiiskii Aero Klub), a decisive association for the development of aviation in Russia, founded in 1908... Ten years after the French Aero Club.

2.1 – Russian aviation under French influence

2.1.1 - A strong French presence from the beginning of the conquest of the air

The conquest of the air is a human adventure that has been shared by several European countries as well as the United States of America. If the first men were able to leave the cow floor at the end of the 18th century in balloons, it was not until the end of the 19th century that a number of technological advances were made with the experiments of several inventors who created gliders that they tried to fly without success with steam engines. At the beginning of the 1900's, petroleum engines allowed the first machines to go higher than the simple jump. At the beginning of the 1910s, this era of pioneers came to an end and the first aircraft manufacturers appeared, producing mass-produced aircraft to meet the first public orders from European armies.

Like all European countries, Russia had its share of enthusiasts and pioneers who participated in the conquest of the air. But the vast empire of the Tsars, with a population of more than 130 million, suffered from a number of handicaps, starting with the fact that it was a conservative autocracy where serfdom was not abolished until 1861 and where the social situation was explosive with the emergence of an urban proletariat, the breeding ground for ever-growing revolutionary unrest. The sharing of power with an elected assembly, the Duma, conceded after the 1905 revolution, was in fact only a façade and the Tsar retained most of the power. In this vast empire with very limited communication channels and where the whole society was under the surveillance of the political police, there were many obstacles to innovation and the development of industry was very limited. As a result, the conquest of the air in the Russian

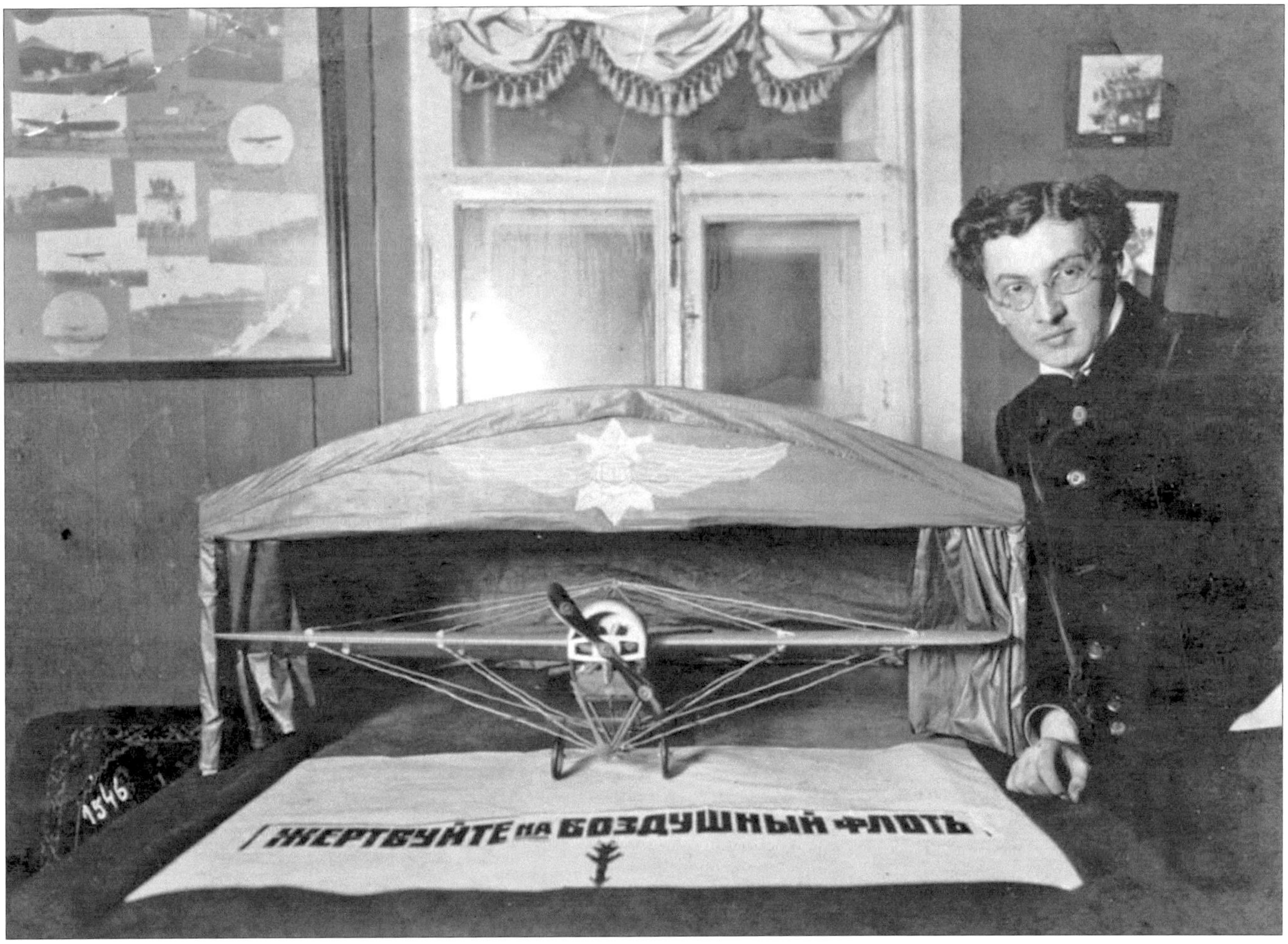

Above: The IVAK organized a great national subscription to provide Russia with a military aviation. The logo of the association is on the front of the hangar, which houses a model of Nieuport IV, the French monoplane which received the largest public order from the Russian state in 1912. (DR)

Empire was always carried out with a certain delay, and always with the help of the country which took the lead: France.

The first man (and woman) to take to the air in a documented way in Russia was a French couple, the aeronauts André-Jacques Garnerin and his wife Jeanne Labrosse who made a balloon ascent on July 2, 1803 in the courtyard of the cadet school in St. Petersburg, in front of Tsar Alexander I and his family. Other Russian aeronauts followed in their footsteps… In 1878, the first Russian aeronautical circle was founded in St. Petersburg, while the Tsar's army and navy began to take an interest in captive balloons. Several pioneers and inventors experimented with heavier-than-air balloons, starting with naval officer Alexander Mojaïski who, in 1884, with the help of an inclined ramp and a steam engine, managed to make a quick jump on an aircraft of his own design, which was destroyed in the attempt. At the end of the first decade of the 20th century, while the first daisy cutters were taking to the air all over Europe, little progress was made in Russia to the point that in December 1907, Vasily Korn, who a few months later would be one of the founders of the All-Russian Aero Club, wrote in an article: "*How can we explain the fact that we have only developed [our science of flight] to a level comparable to that of France at the time of the Montgolfier brothers? Is it really true that the Russian genius exists in such an embryonic state that it is impossible for us to establish something of even the weakest nature? Is it really so difficult for us to measure up to Europe, or even to the whole cultured world?*"

However, there is an awareness of the political staff. Tsar Nicholas II himself gave his blessing to the creation of the Russian Aero Club, which could add the title "Imperial" to its name and became the Imperatorskii Vserossiiskii Aero Klub (IVAK) on January 29, 1908, setting up its headquarters in St. Petersburg and federating all the other aeroclubs that opened in the country. It was at this time that the first military credits for aeronautics arrived and that the first airships were ordered from France. Russian officials noted the much more interesting potential of airplanes and Russian officers attended with interest the flight demonstrations made by Wilbur Wright in France in June 1908. In January 1909 Tsar Nicholas II authorized the IVAK to launch a public subscription for the creation of a military aviation, whose subscription committee was chaired by his cousin,

Above: The Dux factory in Moscow, a bicycle construction company converted to aeronautical production at the beginning of the 1910's, produced mostly French aircraft under license. The Russian aeronautical industry, the weakest of all the belligerents, produced only about 5,600 aircraft from 1914 to 1917 (of which 1,600 were produced by Dux, the largest factory), while 27,500 aircraft were produced in the same period in France, 22,250 in British factories, and 27,385 in Germany... The situation for engines was even worse, with between 1,000 and 1,500 engines produced in local factories, depending on the source. In these conditions, the whole Russian aeronautics industry depends on imports, mostly from France. (Alan Dukota collection)

Below: The First Russian designed aircraft, the Rossiya B, produced during summer 1910 by the Shchetinin company, whose logo can be seen on the tail fin. The aircraft looks very similar to the Blériot XI.

Above: Nieuport IV monoplanes decorated with Russian roundels at the beginning of the war. With nearly 300 locally produced or imported aircraft, the Nieuport IV was the first aircraft to serve in the Russian Imperial Air Force (Coll Michel Bénichou)

Grand Duke Alexander Mikhailovich. Trained as a naval officer, he was passionate about the nascent aviation. On July 25, 1909, he was on vacation in Biarritz and read in the newspapers about Louis Blériot's exploit of crossing the English Channel. He rushed to Paris to meet and congratulate the aviator, as well as to examine his airplane.

It was not until September 27, 1909 that an airplane took off for the first time in Russia when the French aviator Georges Legagneux took his biplane Voisin into the air on a racecourse near Moscow in front of an enthusiastic crowd. It was not until March 21, 1910 that the first Russian aviator, Mikhail Nikiforovich Efimov, took to the skies in Odessa aboard a Farman IV that he had learned to fly in France a few months earlier. In May 1910, the first major aviation meeting in Russia was held in St. Petersburg, where French and Russian aviators demonstrated their flights for several days in the presence of the Tsar and many government officials.

The first Russian military aviators were trained shortly afterwards, while orders for aircraft multiplied in France under the supervision of Grand Duke Alexander Mikhailovich who visited all the French factories and flying schools. In 1911, the Nieuport company won a very large Russian order for 150 examples of its Nieuport IV monoplane, which was the contract of the century at the time and allowed the company, which was in financial difficulty, to stay afloat. The fact remains that the Nieuport company was not the only one to benefit from Russian orders, since this was also the case for Blériot, Morane-Saulnier, Farman, Voisin, and Deperdussin - the entire French aeronautical industry benefited from the Grand Duke's chequebook.

2.1.2 - A Russian aeronautical industry under French influence

However, not all the aircraft ordered were produced in France because the Russian authorities wanted to produce aircraft on their soil and negotiated production licenses with a few Russian industrial machine-building companies.

In 1909 the Shchetinin company was created in Saint Petersburg by the initiative of IVAK members. This company claimed to be the first Russian aeronautical company and in the summer of 1910 it produced the first Russian-designed aircraft, the Rossiya-A biplane and the Rossiya-B monoplane, which was neither more nor less than a copy of the Blériot XI. Directed by its founder Sergei Sergueïevitch Shchetinin, the company then produced several French aircraft under license and its chief engineer, the Ukrainian Dmitry Pavlovitch Grigorovitch, specialized in seaplanes a little by chance by repairing a Donnet-Leveque seaplane in 1913 – the first of his own designs flying in 1914.

At the same time, in Moscow, the Dux company, founded in 1894 and which until then had been producing bicycles and motorcycles, opened an aeronautical workshop. In the summer of 1910 it opened an aeronautical workshop and assembled its first six aircraft, Farman IV. The production line grew steadily and by 1914 the factory had up to 300 workers, producing almost all types of French aircraft ordered under license.

In 1912, another aviation factory was opened in Riga, belonging to the Russko-Baltiisky Vagonny Zaved (R-BVZ), a German-owned company producing railroad carriages, agricultural machinery, and automobiles (it was the only one to produce them in Russia). Under the initiative of its president, the former high imperial official Mikhail Vladimirovitch Shidlovsky, it diversified and opened an aeronautical production line in Riga. The R-BVZ first produced French aircraft under license, but Shidlovsky wanted to open his own aeronautical design office and entrusted it to a promising young engineer of Ukrainian origin, Igor Sikorsky, who had met and corresponded with the French pioneer Ferdinand Ferber and who had distinguished himself by producing several experimental

Above: Another French aircraft serving in large numbers in Russian units at the beginning of the war, this Henri Farman HF 20. (Coll Alan Dukota)

aircraft in 1911. Having significant technical means at his disposal, Sikorsky gave free rein to his creativity and produced several monoplane or biplane aircraft, some of which were commercialized, particularly the S-10 floatplane sold in small series to the Russian Imperial Navy. In 1913, he distinguished himself by designing the largest aircraft in the world at the time, the Sikorsky "Grand", a four-engine transport plane that gave birth to the Ilya Murometz bomber in 1914. At the dawn of the First World War, the R-BVZ was the only Russian company to produce aircraft of national design.

A little later, the Anatra company opened in Ukraine, founded in 1913 by the businessman Artur Antonovich Anatra, who set up an aeronautical factory in Odessa. He first built a few Farman IV training aircraft, then started to produce Nieuport biplanes, Voisins, and Moranes. Under the technical direction of the French engineer Elisée Alfred Descamps, the company went on to produce during the war aircraft of its own design, the Anade (Gnome engine), Anasal (Salmson engine), and Anadis (Hispano engine) two-seater biplanes.

The last important company was the factory founded in April 1914 in St. Petersburg by the pilot Vladimir Alexandrovich Lebedev, which produced a few French aircraft under license (Deperdussin, Voisin L, Morane L, and FBA seaplanes) and then during the war produced a few models of its own design, under the sign Lebed (swan, in Russian), which were mainly copies of captured German aircraft (the Lebed VII was a copy of the British Sopwith Tabloid, the XII was a copy of the German Albatross C).

If Russia had an aeronautical industry at the dawn of the war thanks to these five manufacturers as well as several other small companies more artisanal, it was handicapped by a big bottleneck which constituted the production of engines, because there was quite simply no national company ready to convert its production to carry out engines of planes. The Russian authorities then called on French entrepreneurs, starting with Mr. Louis Seguin, the director of the Gnome company, who was invited to St. Petersburg in February 1912 with a French military-commercial delegation. He opened a local subsidiary and entrusted the French engineer Emile Germinet with the construction of an engine factory in Moscow, which was to remain the only one in Russia until 1914, when a subsidiary of the French engine manufacturer Salmson opened in the same city, and then, at the beginning of the war, another factory in Petrograd, a subsidiary of the R-BVZ, to produce copies of French, British, and German engines intended for Murometz bombers. During the war, these three factories would ensure almost the totality of the meager Russian production which would be dramatically insufficient to cover the needs of the tsarist military aeronautics, which would largely depend on imports.

Above: One of the few aircraft of national design, the four-engine Ilya Murometz bomber had a difficult start in an autonomous Imperial Air Force squadron, reporting directly to the Russian Army Staff, the Stavka. (Coll Colin Owers)

2.1.3 - The test of war: a mediocre performance of the air force

At the time of mobilization, the Russian air force was very powerful on paper, with 30 squadrons, one for each army corps (Korpusnoi Aviatsionniy Otryad - KAO), plus a squadron reserved for the 3rd Army staff, as well as eight so-called fortress squadrons for the strongholds located in Poland, which was then attached to the Russian Empire. This represented a theoretical force of 263 aircraft, although the number of personnel planned by the General Staff was theoretically twice as high. This is still a larger total than that of France, which at the beginning of the war had 138 aircraft in 21 squadrons plus 3 light cavalry squadrons, the total rising to 274 if one adds the reserve aircraft and those of the naval aviation. But this numerical comparison has its limits if we take into account the fact that many of the Russian aircraft were not in flying condition, and that there were only 231 military pilots to fly them, whereas 492 military pilot licences had been awarded in France by July 31, 1914, which also had a good pool of civilian pilots who would quickly increase this number to 614 by the end of September.

The ordeal of the war quickly demonstrated the incompetence of the Russian General Staff (known as the "Stavka", the command tent), which the Tsar entrusted to his uncle, Grand Duke Nicolas Nikolaïevitch of Russia. There was no unified command of the air force, whose squadrons were placed directly under the orders of the generals commanding the army corps or the fortresses. Many of them do not seem to pay much attention to the observations made by their airmen...

It was on the northern front that the Russian army would experience a terrible fiasco. After the mobilization phase, two Russian armies launched an assault on East Prussia, which was defended by the 8th German army led by General Paul von Hindenburg and his chief of staff, General Erich Ludendorff, and which was more than twice as small in number. In mid-August 1914, the Russian 1st Army led by General Rennekampf (who had 4 squadrons) arrived from the Baltic States and attacked westwards, forcing the German army to retreat. Meanwhile, General Samsonov's 2nd Russian Army (with 5 squadrons, 1st, 13th, 15th, 21st, and 23rd KAO) was positioned in the Polish salient and moved northwards. The Germans, who were listening to the unencrypted wireless communications of the Russian armies and made perfect use of their air force, were able to spot the enemy positions and move their troops accordingly. They first annihilated the 2nd Russian army during the battle of Tannenberg between 26 and 30 August 1914, General Samsonov committing suicide to avoid a humiliating capture. The 13th KAO was annihilated during the battle and all other squadrons of the 2nd army lost most of their equipment.

On August 27, 1914, the Stavka decided to establish a general command of the air force, which it split in two, entrusting the direction of the southern front (facing the Austro-Hungarian armies) to Grand Duke Alexander Mikhailovich, and the northern front to Baron Alexander Vasilyevich von Kaulbars, a 70-year-old cavalry general, who took command in the midst of the turmoil of the battle of Tannenberg. The victorious German 8th army then turned against Rennekampf's 1st Russian army, which had given no support to Samsonov, and defeated it at the Battle of the Mazures Lakes between September 7 and 14. At no time was the Russian air force, which according to German reports did not venture beyond the front line, able to effectively inform

Above: The first victorious pilot and future ace of Russian aviation, Poruchik (Lieutenant) Alexander Kazakov, a pilot with the 4th KAO, succeeded on March 31, 1915 in shooting down an enemy two-seater aboard his Morane G No. 316, which he had equipped with a grappling hook fitted with explosives. He actually hit his target more or less unintentionally and left his own landing gear behind, but managed to return with his damaged aircraft. Kazakov later became the head of the main fighter group of the Russian Air Force and crossed the path of French pilots who came to Russia several times. (DR)

the command about the position of enemy or even friendly troops to such an extent that in order to inform the high command about the position of Rennekampf, the old general von Kaulbars himself left on a Farman 22 piloted by Shtabs Kapitan A.K. Perepechin, from the Kovno fortress squadron. The plane flipped over on take-off because of the wet ground and the general came out with several broken ribs… The incident did not help to raise the low esteem that Grand Duke Nikolai and the Stavka have for the aviators.

However, there was one general who understood their usefulness, General Broussilov, who commanded the 8th Russian army, the southernmost of the four armies of the south-western front facing the Austro-Hungarian troops. The latter did not have the same calibre as their German ally and were defeated by the Russian troops, who forced them to retreat towards the Carpathians. In this war of movement, Broussilov, undoubtedly the most competent general in the Russian army, made maximum use of his airmen, but quickly realized their technical limitations to such an extent that on September 13, 1914, he visited the Grand Duke Alexander Mikhailovich to ask for reinforcements: "At present, I find myself short of the aircraft that I vitally need for reconnaissance, which leaves me with great difficulty in knowing the enemy's positions. I humbly beg your royal highness to render to my army the services of the Farman and Nieuport. Aerial reconnaissance is indispensable to me.

The first five months of the war in 1914 began very badly for the Russian army and its air force. As soon as winter was over, hostilities resumed for the year 1915 that was to prove even more complicated. On January 18, 1915, the Tsar finally decided to create a unified command of the air force, which he entrusted to his cousin Grand Duke Alexander Mikhailovich, who was appointed Inspector General. However, this was only theoretical and did not reflect the eminently pusillanimous character of the Tsar under the influence of his wife, herself under the thumb of the monk Rasputin.

Indeed, on October 15, 1914, in front of the poor operational results carried out by the four-engine Ilya Murometz bombers of which 10 specimens were ordered at the beginning of the conflict, the ministry of the war decided to cancel its order of a second 22 specimens to the company R-BVZ, which puts it in financial difficulties. It was to count without the dynamism of its president, Mr. V. Shidlovskiy, which decided to go to plead directly its cause to the Stavka to maintain the order. He pleaded with the Tsar and Grand Duke Nikolai that the bombers, rather than being dispersed individually, should be grouped into a large autonomous strategic unit (called "heavy aviation") of which he would take command and which would be accountable only to the Stavka… His proposal was accepted and the command of the bombers thus escaped the Grand Duke Alexander who commanded the so-called "light aviation" and who, out of jealousy, did not fail to put obstacles in Shidlovskiy's way, in particular by removing personnel from the ground staff,

Above: The most famous French aviator in Russia, Alphonse Poirée, who flew his personal Farman in Moscow before the war, joined the Russian Air Force and took part in the Battle of Tannenberg. He survived the war and, having become a test pilot, died in 1922 in a plane crash aboard a Caudron (DR)

Above: A mysterious adventurer, Jean de Voyod is French by his father, an opera singer, and Russian by his mother. He served in the Russian air force in the 16th KAO, then at the Odessa school. During the civil war, he was enrolled in the Bolshevik air force but joined the White Army of Admiral Kolchak with his aircraft. He then went into exile in China where he put himself at the service of one of the local warlords, building him an aircraft.

which was detrimental to the maintenance and operational capacity of the four-engine aircraft whose performance allowed them to carry out long distance reconnaissance and to carry a substantial bomb load.

The latter, if better used, would probably have been very useful for the Russian army to bombard more efficiently the stations and communication nodes used by the German army, which undertook a vast offensive at the beginning of 1915 to push back the Russian troops and put them out of action. The German offensive was handicapped by the lack of roads, making it difficult to move the masses along a railroad line – hence the usefulness of the Murometz bombers. But the German thrust proved to be effective and during 1915 the Russian army was forced to retreat eastward and abandon all of Poland, whose strongholds fell one after the other under the blows of the large-caliber howitzers of the Kaiser's army. The Austro-Hungarian troops, advised by Germany, took back all the ground lost to the Russians during the 1914 offensive and pushed them back beyond their border. The Stavka was unable to organize a solid resistance, simply because of the bankruptcy of the Russian economy, badly managed by the bureaucracy, which could not provide its army with the supplies to sustain a long war. In August 1915, the Tsar dismissed Grand Duke Nicholas, who decided to assume command of the army himself, although in fact this responsibility fell to his chief of general staff, General Mikhail Alekseyev.

In the rear, the social situation was explosive and numerous strikes broke out in the factories. In June 1915, the French ambassador, Maurice Paléologue, reported his meeting with the Russian industrialist Alexis Putilov, who told him that "the days of tsarism were numbered, that revolution was inevitable, that the crisis in the supply of munitions was not only a technical problem but that the question of manpower had to be solved, and that finally the

Above: Another original destiny is that of Marcel Plat, a Frenchman of mixed race who spent his adolescence in Russia and then worked as a mechanic and driver in the R-BVZ factory manufacturing Ilya Murometz aircraft. He served during the war on these heavy bombers as a gunner, and returned to France at the beginning of the civil war.

Above: A descendant of a French or Swiss family who had come to live in Russia, the polkovnik (colonel) Léon Ducimetière, of Russian nationality, served as an observer and then as a pilot, and in 1916 he became the deputy of Grand Duke Alexander, who was in charge of the Imperial Aviation. After the fall of the Tsar, he participated in the attempted coup d'état of General Lavr Kornilov on August 27, 1917, then served in the White Armies, and took refuge in China after the civil war. (Coll August Blume)

entire administrative system of Russia had to be changed from top to bottom.

This chaotic situation was particularly detrimental to Russian aviation, which unlike the aviation of other belligerent nations, saw its workforce decrease during the year 1915... The Tsarist bureaucracy did not understand the need to mobilize the industrial potential before May 1915, local production was anemic, especially for engines. The Gnome factory in Moscow could only produce 17 engines per month, when it was not the victim of supply shortages at its subcontractors. As for the aircraft produced in Russian factories, many of them were difficult to use at the front because of multiple defects, as was the case with the Voisin produced in the Shchetinin factory in St. Petersburg, which were the subject of a report from the command, indicating that the airmen refused to fly on them. The vast majority of losses suffered by Russian airmen and their aircraft have absolutely nothing to do with enemy activity, but were caused by the deficiencies of their own aircraft, or even by the flak fire of their own troops...

The fighter was still very little developed, both on the German and Russian sides, and aerial victories were rare. The first one won by a Russian aircraft was the work of the Chtabs-Kapitan Piotr Nesterov who, on September 8, 1914, tried to destroy an Austrian Albatros B.II with its landing gear: he succeeded at the cost of his life because the two tangled planes crashed to the ground. It was in fact the next day that the first claim of victory took place, made by a crew of the Grodno fortress squadron (Poruchik Aleksei I. Semenov at the controls, and Kornet Nikolaev as an observer) firing on a German aircraft between Lotzen and Rastenburg (now Giżycko and Kętrzyn, respectively, in Poland), which was then targeted by flak and reportedly forced to land near a lake where it was destroyed by artillery, with no identification of the aircraft in the German records.

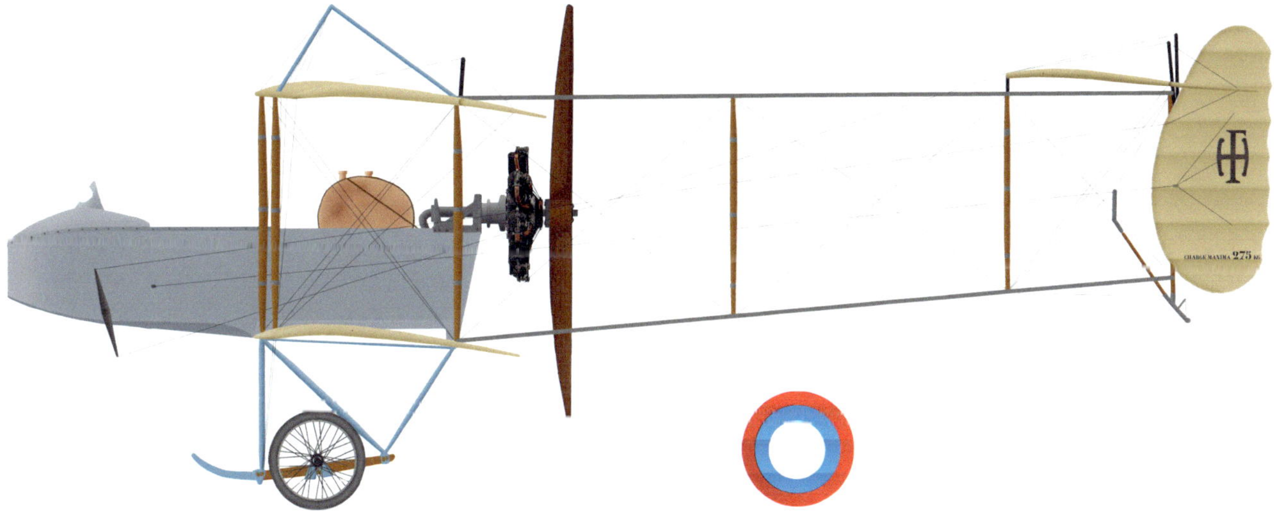

Henri Farman HF 20 (unknown no.) flown by Unteroffitser (adjutant) Alphonse Poiré, assigned to the squadron reserved for the command of the 2nd Russian Army (2nd AAO) in January 1915.

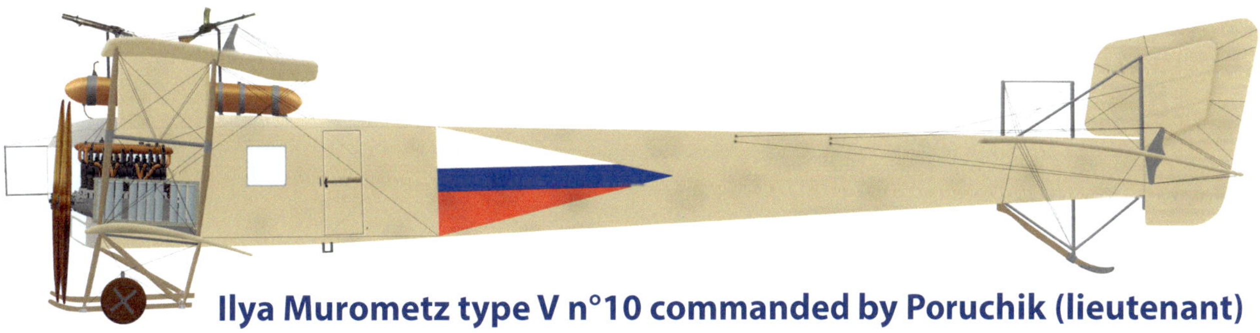

Ilya Murometz type V n°10 commanded by Poruchik (lieutenant) A. M. Konstenchik and on which served Feldwebel (sergeant-major) Marcel Plat, making on April 8, 2,3 and 26, 1916 three flights on board the aircraft whose precise registration possibly painted on the fuselage is not known.

The first truly indisputable victory of Russian aviation was in fact achieved on March 31, 1915 by the Poruchik Alexander Kazakov, the future ace of Russian aviation, who alone aboard his Morane G monoplane n°316 of the 4 KAO attacked a German two-seater using a grappling hook equipped with explosives that he left hanging under his aircraft. He miscalculated his attack and hit the German aircraft with his landing gear, which crashed; Kazakov managed to land and destroy his Morane. Despite the absence of fixed armament firing in the front axis, Russian pilots made 19 claims during the years 1914 and 1915 (not counting the Nesterov collision). Due to the lack of certification process, we can estimate that six of them correspond to real victories. Knowing that at the same period, the French aviation which from spring 1915 organized fighter squadrons realize 128 claims whose authorities recognize them 66 homologations.

However, hunting was only secondary to the work that the air force had to carry out, whose primary purpose was reconnaissance and artillery adjustment. However, for the first, the quality of the aerial photographs brought back by the Russian aviators left much to be desired, due to the lack of adapted cameras, when they had them. As for the artillery adjustment, the question is without appeal since they did not do any!

2.1.4 - Several French airmen served in the Russian air force at the beginning of the war

The delay of Russian aviation in the conduct of the war was so obvious that, on the instructions of Grand Duke Alexander, Colonel Ignatieff, the Russian military attaché to the French Grand Headquarters, brought a letter from Grand Duke Alexander on January 20, 1916, requesting that a French military aviation mission be sent.

The services of the French General Staff thought about it... They had an idea of the task thanks to the presence in France of the Unteroffitser (Adjutant) Alphonse Poiré, a French civilian pilot who was in Russia at the time of the declaration of war, and who had joined the Russian Air Force with his own Farman HF 20. Assigned to the 2nd Russian Army reconstituted in Poland after the defeat at Tannenberg, he served in the 21st KAO, 23rd KAO, and then in the 2nd AAO when it was created in January 1915, carrying out numerous reconnaissance and bombing missions, flying more than any other Russian aviator, and receiving the four classes of the Knight's Cross of St. George. He even claimed one of the very first victories of Russian aviation on March 10, 1915 by flying as an observer in an aircraft piloted by the Poruchik S.K. Shebalin, and by shooting down a German plane and forcing it to land in the Lowicz region (west of Warsaw).

Like Poiré, three other Frenchmen also served in the Russian Air Force, as they were in Russia when war was declared and could not return to France after mobilization. Jean de Voyod, a pre-war French civilian pilot, served as a pilot instructor at the Gachina school in St. Petersburg and was transferred to the fighter squadron in 1916, receiving the St. George Cross 1st class. Another Frenchman surprised in Russia by the declaration of war, Henri Bourrier, was a French teacher who joined the Russian army and served in 1915 as a mechanic in the 18th KAO, then became a pilot and commander of the 23rd KAO, receiving the St. George Cross 2nd class. Finally, let us mention the unusual fate of a Frenchman with black skin, Marcel Plat, son of a French maid living in Russia, who became a driver-mechanic at the R-BVZ in Riga before the war; he joined the Russian army in 1915 and served as a mechanic and then as a machine-gunner in the EVK on Ilya Murometz, where he distinguished himself by receiving the St. George Cross 3rd class.

Let us also mention a perfectly French-speaking officer, Léon Pavlovitch Ducimetière, probably descended from a family of Swiss Romans settled in Russia as early as the 18th century and who served with the rank of kapitan at the staff of the 2nd Russian army in Poland where he crossed paths with Poiré at the 2nd AAO at the beginning of 1915. As an observer, he flew a few aerial missions and on March 6, 1915, aboard the Voisin LA n°42 piloted by the Poruchik Yevgrav Kruten, he shot down a German plane in the German lines between Sochaczew and Rawka (west of Warsaw), which he certified as having been shot down, even though neither he nor his pilot were able to see the crash site. But his position as a staff officer of the 2nd army was enough to certify the victory and to receive a decoration... As a pilot, he was promoted to the rank of polkovnik serving on the staff of Grand Duke Alexander in 1916.

Poiré was the only one of these Frenchmen to be recognized and decorated by the French government, which awarded him the military medal on April 18, 1915. In December 1915 and January 1916 he was on leave in France to purchase aircraft for the Russian government and most likely his testimony on the state of the Russian air force was collected and analyzed by the French military hierarchy, which also had reports from its military attaché at the Stavka.

Below: Mostly dressed in civilian clothes, pilots and observers of the French mission of Major Bergé pose with Russian aviation officers during their visit to Petrograd in early May 1916. 1: Henri Millet (observer), 2: Henri Moutach (pilot), 3: Marcel Paillette (pilot), 4: Marc Bonnier (pilot, killed on August 9, 1916), 5: Marc Partreau (pilot), 6: Joseph Ribourt (observer), 7: Marc Coppey (observer), 8: Jean de Lubersac (observer), 9: Emile Stribick (pilot), 10: François Méritel (pilot, killed on June 12, 1916), 11: unknown, 12: Henri Laurent (observer, killed in action on September 5, 1916), 13 : Eduard Pulpe (pilot, killed in action on August 2, 1916), 14 : commandant Paul Berger (mission leader), 15 : Louis de Lignac (pilot, killed in action on July 7, 1917), 16 : Jean Cotier (observer, captured on June 10, 1916), 17 : Léon Boittiaux (observer), 18 : Raymond Berné (observer, killed in action on March 2, 1917). (Coll Albin Denis)

2.2 – The Berger mission

2.2.1 – The arrival of the Berger mission in Russia

The January 20th request to send an aeronautical mission was the subject of several exchanges between French military and political leaders. The principle was accepted in exchange for the Russians sending an infantry brigade to the French front.

Then, on April 11, 1916, the French Grand Quartier Général defined the military mission, which would be composed of 10 pilots and 10 observers. In addition to these 20 pilots, there were also the pilots' personal mechanics, forming a small troop of 30 Frenchmen who embarked in Brest under the command of a mission leader, whose designation and responsibilities were not without difficulty. In the exchange of telegrams, Grand Duke Alexander refused to allow the airmen to be placed under a French military command. The French GQG insisted on having access to the highest level of the Russian command and engaged in a diplomatic acrobatics:

"The French GQG insisted that an officer be designated as head of mission to ensure unity of doctrine in the employment of these airmen. The Russian government did not accept the official designation of a head of mission, but requested that an officer be attached to the person of Grand Duke Alexander, inspector of Russian aviation. The functions of this officer, as indicated in the telegram of the French military attaché, will consist in studying the organization of the Russian aviation and especially to pass inspections on the front. In order to have the necessary influence for the success of the mission, that is:

1- To supervise the work of the French pilots and observers, and to exercise the real command of them, although in an unofficial way.

2- To direct the Russian aviation towards better conditions of organization and use, by a constant personal action with the Grand Duke and his entourage.

It is essential that the air officer seconded to Russia be a battalion commander.

After some negotiations, the head of the aeronautical mission was appointed in the person of Major Paul Henri Berger, an officer from the infantry. He was to be supported by Colonel Henri Janin, who was promoted to temporary general on April 20, 1916 and appointed head of the French military mission in Russia to the Stavka.

The French soldiers reached the country by sea on the auxiliary cruiser "*La Champagne*" between the 20th and the 25th of April 1916, disembarking at the port of Arkhangelsk. All the sailors who were not already officers were promoted to second lieutenant on a temporary basis. The ten pilots included a wide range of French aviation specialties: fighter pilots for Eduard Pulpe (3 victories at N23) and Henri Moutach (1 victory at N26), reconnaissance and artillery adjustment pilots for Emile Stribick (who also won two victories), Louis de Lignac, and Maurice Bonnaud. The other five were Voisin pilots from the Bombardment Groups: Aimé

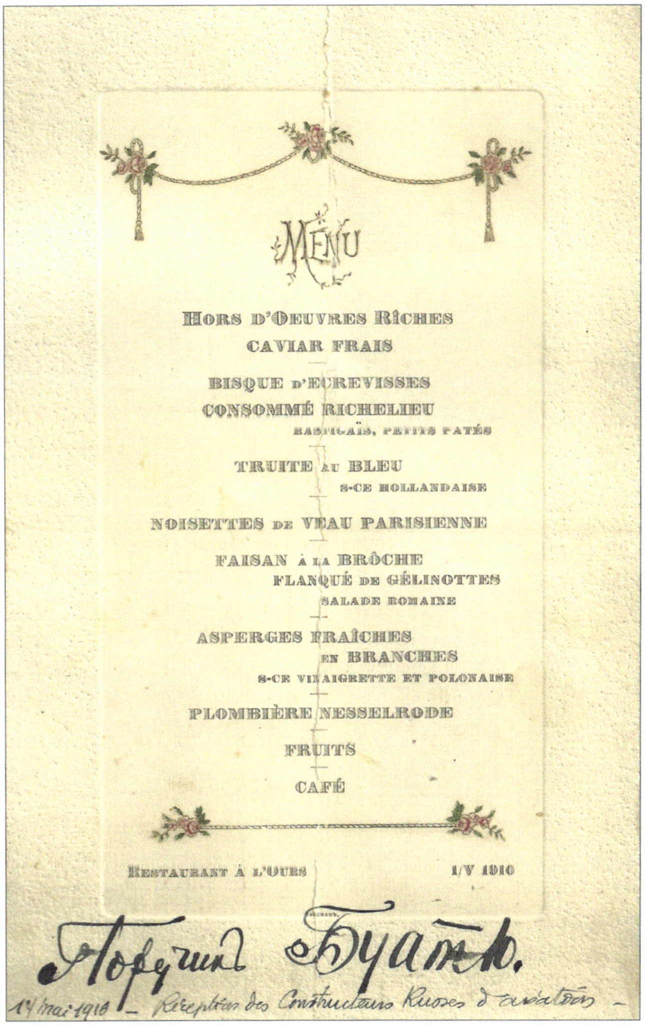

Above: The aeronautical industry of Petrograd offers a sumptuous banquet to the pilots and observers of the Berger mission, on May 14, 1916. Among the signatures on the back of the menu are those of the industrialists Sergei Shchetinin and Vladimir Lebedev. (Coll Albin Denis)

Grasset, François Méritel, Marcel Paillette, Marc Partreau and especially Marc Bonnier, a famous pre-war pilot who flew a Paris-Le Cairo link on a Nieuport monoplane in 1913, he also distinguished himself during the war in VB 102 during a raid on Ludwigshafen on May 27, 1915. As for the observers, they were divided between specialists in bombing, aerial photography, and artillery adjustment by radio. It should be noted that two other Frenchmen later joined the mission: a pilot, Jean Robinet, and an observer who later became a pilot, Georges Sausse.

From Arkhangelsk the French mission travelled by train to Petrograd (the new name of St. Petersburg) and received a very warm welcome from the Russian military authorities, who had a Russian uniform made for each Frenchman corresponding to his rank (podporuchik for the most part). On May 8, French pilots and observers were in Moscow at the invitation of the city's aeronautical industrialists (Dux, Mosca, Gnome, Salmson). On May 14 they were

Above: A few days after their arrival in Petrograd, the French aviators of the Berger mission, who all received a Russian uniform, pose in this photo with Grand Duke Alexander (marked with a cross), head of the Russian Imperial Air Force. On his left, having kept his uniform and his French kepi, is commandant Paul Berger. (Coll Albin Denis)

again in Petrograd at the invitation of the local aeronautical industrialists (Shchetinin, R-BVZ, Lebed) and three days later they participated in a new banquet given in their honor by Grand Duke Alexander himself who came to receive them.

2.2.2 – The French in combat during the Broussilov offensive

The French detachment then left for Kiev where pilots and observers were distributed among the Russian squadrons in four separate groups, most of them being assigned to the southwestern front where the Russian army was going to launch a large-scale offensive following an agreement between the nations of the Triple Entente, which had the ambition of carrying out a combined attack with the Franco-British offensive on the Somme and the Italian attack on the Isonzo.

Between the Pripet marshes and the Romanian border, no less than five Russian armies went on the attack, facing mainly Austro-Hungarian troops who would effectively retreat in this offensive that would last until winter. The Stavka entrusted the command of the offensive to the competent general Broussilov, who had already undertaken several reforms in the use of troops, who were much better supplied with ammunition and artillery shells than the previous year. Brusilov was fully convinced of the importance of the air force, which on March 25, 1916 began to reform itself by creating its first three fighter squadrons (IAO, Aviatsionniy Otryad Istrebitelei) and finally experimenting with artillery adjustment in April. The whole problem of Russian aviation remained the supply of aircraft...

On January 13, 1916, the Stavka counted a total of 58 squadrons, with 311 aircraft at the front, including 25 Nieuport 10, 98 Morane Parasols, 11 Voisin LAS, 31 Farman, 6 Deperdussin, 31 captured enemy aircraft, and 5 miscellaneous aircraft. To this total we must add 76 aircraft in reserve in the fleets and 112 under repair, i.e. 499 aircraft. This is an inventory of outdated aircraft, roughly equivalent in number to the forty or so German squadrons (one third of the Luftstreitkräfte's strength) and the nine Austrian squadrons then present on the Russian front. By way of comparison, in February 1916, France had 937 aircraft at the front and probably a little less than 200 others in the Camp Retranché de Paris and on the Eastern Front, i.e. between three and four times more personnel. The same is true for Germany, which, according to its inventory, has 1,301 aircraft at the front and in its reserves... Russia had in fact increased its air force strength.

When the Broussilov offensive began on June 4, 1916, French pilots and observers were all posted in Russian squadrons and took part in the first battles of this vast offensive, which would not end until the winter.

The culture shock was harsh for the French, who discovered units with rudimentary comfort: no Bessonneau hangars or Adrian huts in the frontline squadrons, but

Above: The French pilots and observers of the Berger mission are distributed among the Russian squadrons, where they discover very rustic living conditions. Second Lieutenant Boittiaux photographed this Voisin LAS stuck in what seems to be a barely thawed pond. (Léon Boittiaux, coll. Albin Denis)

Above: Another surprise for the French aviators of the Berger mission, the discovery of outdated equipment, such as this Morane parasol which was withdrawn from French units at the end of 1915 and which is still in service nearly a year later in Russia (Léon Boittiaux, coll. Albin Denis)

Left: In this photo, the pilot Marc Partreau poses standing behind the observers Léon Boittiaux (left) and Augustin Chabert (center), with a certain Constantin on the right who is perhaps the mechanic Constant Philippon belonging to the French mission. Chabert's private correspondence, intercepted by the Russian censors, will be the cause of a serious diplomatic incident and will lead to his dismissal to France - the Grand Duke Alexander not liking that his country be qualified as "primitive" (Coll Albin Denis)

aircraft housed in rudimentary tents and pilots staying in isba built on the edge of the field or in the homes of local residents. The equipment was late compared to the French squadrons, the Russian-made equipment did not inspire much confidence, and the logistic chain was deficient, limiting the flights due to various supply problems. It was not uncommon to see several different types in a squadron. Russia was also a dictatorship and the Tsar's police were never far away to keep an eye on everyone. The social tensions of Russian society appearred quite cruelly to them, with tensions felt against the staff officers from the aristocracy.

Second Lieutenant Emile Stribick, a very experienced pilot, recounted his experience on the Russian front in these words:

Below: During the summer of 1916, photo of a group of Russian airmen on the Belarusian front, with 3 Frenchmen of the Berger mission among them: Raymond Berné (observer at the 2nd KAO, 2nd left), Léon Boittiaux (observer at the 27th KAO, 9th left) and François Méritel, (pilot at the 27th KAO, 10th left) (Coll. Albin Denis)

Above: However, the first losses in combat dissipated the unease caused by the Chabert affair. On June 13, 1916, the funeral of the pilot François Méritel and the mechanic Maurice Bouvrier, accidentally killed the day before on board a Morane Parasol, was organized in the Cathedral of Saint Mary in Minsk and gathered a large crowd (Coll. Albin Denis)

"Our life on the French front was sweet compared to what awaited us on the Russian front. As soon as we arrived, contrary to what we had hoped for, we were all separated. Ignorance of the language, slow communications, lack of all the elements necessary for useful air warfare work placed each of us in complete isolation. Russian pilots, who on the whole are of a much inferior quality, have no idea of the special training that modern warfare requires. Reconnaissance aviation, photography aviation, artillery adjustment aviation, and even bombing and fighter aviation require a special science and cooperation with the years to which the Russian aviation has always been alien.

The terrain, especially for me who was assigned to the Carpathian armies, caused me intense amazement. The country, all mountains, offered to the aviators only very rare plains, far from each other. The only land that could be used, because of the need for supplies, was that which was close to the communication routes. Now there the roads – and how rudimentary during the rainy season and during the winter – are excessively rare and the railroads are an exception. As a result, our airfield, from which we left for our expeditions on the lines, was 80 kilometers from the trenches.

Each of my trips involved at least 200 kilometers, in a freezing cold that we cannot imagine here, over high mountains, steep valleys, massive forests, and deadly precipices. I mention only for the record the swamps and the snow plains with their treacherous and dangerous nature. As for the landmarks, the list was simple: no seriously, scientifically established map. The first and indispensable task – and, I must say, highly appreciated by the Russian command – was to establish a map. Our colleagues in the Russian Air Force did not know the rigorous mechanism of taking pictures at the same altitude, under the necessary conditions, which would allow to fix the topography of a sector. This was the essential task for me in my army, for Ram in his, for each French airman according to his assignment. We all fulfilled it and I hope that later on it will be known.

The conditions in which we carried out this thankless and painful work, it is easy to imagine them. The cold, the unknown country, were the essential circumstances, but there were also the absence of supplies, the lack of spare parts, the ignorance of the mechanics, the insufficiency of all the aviation personnel. In addition, following the custom born of Russian indolence and negligence, the railroads gave only a derisory output and the gasoline, the engines, the spare parts reached us only with an improbable irregularity. Breakdowns, in such a country and

Above: French pilot Marc Bonnier in Russian uniform during the summer of 1916, with his wife Marthe Roussi whom he married in Paris in 1912 (Coll author)

under such conditions, became tragic. One day I landed in a ravine without being killed, breaking my plane.

I risked crossing the Carpathian Mountains after making makeshift repairs and making artillery adjustments on the enemy at ground level with a winded engine. All my French friends went through the same adventures. We had, moreover, only a natural merit, since our devotion was essential to the success of the operations and it was with our collaboration that General Broussilov and General Kaledine succeeded in capturing three hundred thousand Austro-Germans."

The integration of French pilots and observers is not self-evident. On the one hand, many Russian officers felt that they had no advice to receive from anyone, and on the other hand, the French were exasperated with this society and army that they considered backward. This led to an incident on July 19, 1916 at the 8th ΛΛO, when Second Lieutenant Marc Parteau and his observer, Second Lieutenant Augustin Chabert, refused to fly on a Voisin-Canon which was put at their disposal, but without the 37 mm shells that the Russian logistics had forgotten to send. Their squadron leader proposed to replace the cannon with a machine gun, but the French crew refused because the aircraft would be unbalanced due to a change in weight. The refusal of the two Frenchmen was welcomed by the Russian hierarchy which moreover discovered, via the postal censorship, the correspondences of the Second Lieutenant Chabert who did not take any gloves to express what he thinks of his hosts and even directly questions the military censorship:

"*Miss censorship,*

Be more charitable, why do you withhold all my letters? You are beginning to annoy me. I know that all truth is not good to say, and especially that which stings, and this is why you get angry. You cut with your big scissors, blacken with your big black brushes and make our letters disappear.

I say this especially to you, Russian censors, because the French censors do not protest. Don't see spies everywhere, we are studying your qualities and faults here and we can help you to rectify them.

If we talk about this, on our part it is a form of camaraderie. Above all, do not look for spies where there are none, all the French airmen are good Frenchmen and all of them have proven themselves in France, and therefore in principle they are all your friends, even if like me they do not admire everything that is happening in Russia. Actually, there is not so much to admire here: oh no! If you want to look for spies, look for them among your Red Cross ladies who are at the front and lead a scandalous life with all the officers coming out of the trenches and with all the staff officers. Don't forget that these ladies force you to talk, that they question you all the time, and that they all speak German. I know this although I have never had any relations with them, but I have often spoken with them when they were with Russian comrades, and I know what I am saying. Note that in France all women were kept away from the front line medical units, and for good reason. So take note of all this.

Signed Chabert, French aviator"

Chabert confided in his other letters to his wife that he was bored because he can not fly and, speaking of the Russians, believes that "*We could have done so great things with such beautiful troops, so disciplined, so brave, and so patient. But here, for that you need reins, rulers tending to the same goal, and not people with pockets too deep.*"

The affair went up to the Grand Duke Alexander, who demanded the immediate dismissal of Chabert and Parteau. Major Berger and General Janin were in a quandary and tried to save Parteau, who had not put anything in writing. His Imperial Highness agreed to keep him and change his army, but the staff of the 8th Russian Army wanted to get rid of him "to remove all traces of the incident. The two French airmen left for their country on the first train to Arkhangelsk...

The Chabert affair reveals the sometimes difficult reception given to French airmen by the officers of the Russian squadrons, who consider that they have no advice to receive from foreigners. However, the brotherhood of arms in combat eventually dispelled these misunderstandings: the French were at the heart of the fighting in the Broussilov

Above: An Anatra two-seater produced by the Odessa factory (Coll. Jack Herris)

Right: Anatra two-seater in which Marc Bonnier was killed in Odessa on August 9, 1916, during a test flight. The countersigned report from French pilots indicates that the crash was due to a piloting error, Bonnier having tried to glide to his field instead of landing in an emergency. (Coll Albin Denis)

offensive and suffered serious losses.

As early as June 10, the first week of the offensive, a first loss was to be deplored when Second Lieutenant Jean Cotier, observer at the 32nd KAO (7th Army), was captured by the Germans after his Voisin, piloted by the Praporshchik Bertold Ya. Shvarts, suffered two aircraft attacks and a flak hit. It fell into enemy lines and the French-Russian crew was captured at Monasterzhiska near Podgaitsy, 100 km SE of Lviv.

The next to fall in battle was one of the fighter pilots, *Sous-Lieutenant* Edwards Pulpe, born in Riga of a Lithuanian father and a French mother, who had settled in France just before the war where he obtained his civilian pilot's license. When the war broke out, he joined the Foreign Legion and went into the air force, being assigned to the N 23 squadron from 1915 until the beginning of 1916, where he won three certified victories. Assigned to the 8th fighter squadron (Aviatsionniy Otryad Istrebitelei - AOI) equipped with Nieuport 11s, he fought in the Kovel sector where the German troops organized a vigorous counter-offensive and where aerial encounters were most frequent. On July 1, 1916, he claimed an aerial victory in the sector between Loutsk and Kovel, then, on August 2, he faced three German two-seaters in a very long battle at the end of which he died, hit by three bullets and crashing near Kovel.

Finally, on September 5, 1916, the Nieuport 10 n°711 of the 5th fighter squadron (5th IAO) on the Northern Front (5th Army), piloted by Praporshchik Onisim Petrovich Pankratov and carrying as gunner S/Lt Henri Laurent, was attacked by a German aircraft of FFA 58 and shot down near the Drysvaty lake, killing its Franco-Russian crew.

In addition to these three Frenchmen killed in combat, three others were killed in accidents, confirming the fact that the main danger faced by Russian airmen came from their own aircraft. On June 12, 1916, in the 27th KAO, a Morane parasol flown by Lt François Méritel and his mechanic, Private Maurice Bouvrier, crashed on the Minsk airfield.

Above: S/Lt Léon Boittiaux aboard a Farman F.40 of the 27th KAO on the Russian front during the winter of 1916–1917. He is sitting in the front seat of the observer. (Coll Albin Denis)

Then, on August 9, 1916, the most famous French pilot of the mission, S/Lt Marc Bonnier, killed himself by crashing in a spin on an Anasal D (n°190) at the Anatra factory in Odessa where he had been assigned as a pilot receiver. The report of the accident, countersigned by the French pilot Jean Robinet, who was also the pilot in charge of the factory, attributed the accident to a piloting error while he was gliding back to the field, and not to a defect of the aircraft. Major Berger noted: "*S/Lt Bonnier was one of our oldest and best pilots. He could have landed before reaching the airfield in a field, at the risk of breaking the aircraft. For his own sake, he preferred to straighten up and try to prolong his glide, hence the loss of speed.*

Announced by the French newspapers, Bonnier's death and even more so that of his French comrades who had fallen at the front gave rise to solemn funerals organized by the Russian authorities, especially for Méritel and Bouvrier, whose coffins were transported to Saint Mary's Cathedral in Minsk on the fuselage of a Morane, in front of a crowd of soldiers and civilians. These successive tragedies created a certain fraternity of arms and contributed greatly to the integration of French airmen in Russian units.

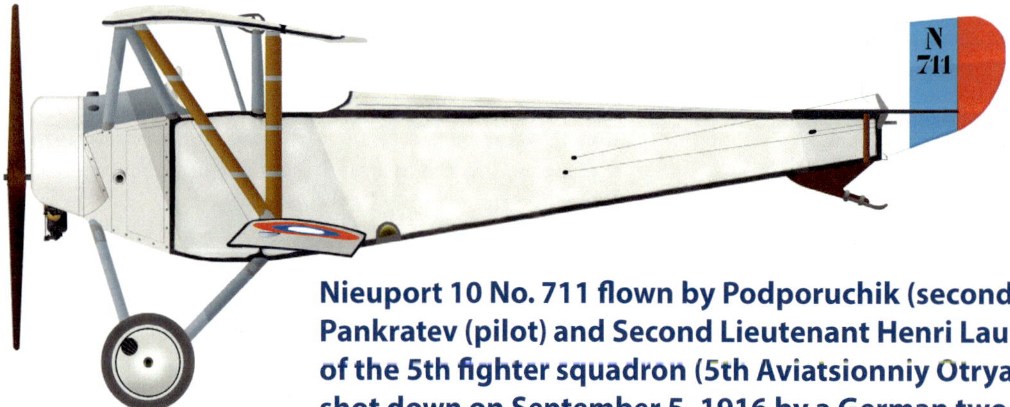

Nieuport 10 No. 711 flown by Podporuchik (second lieutenant) O.P. Pankratev (pilot) and Second Lieutenant Henri Laurent (gunner), of the 5th fighter squadron (5th Aviatsionniy Otryad Istrebitelei), shot down on September 5, 1916 by a German two-seater near Lake Drysvaty in the Dvinsk region.

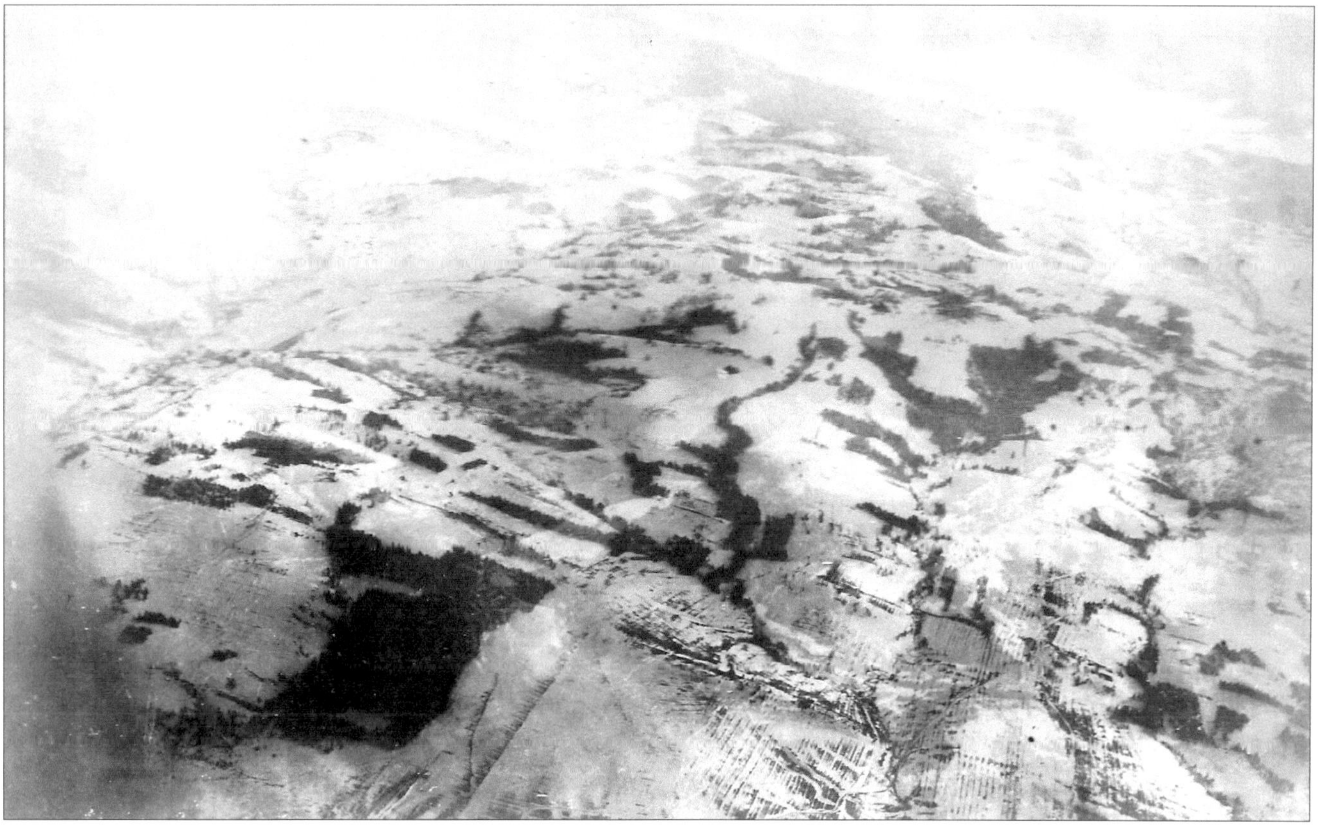

Above: The snowy hills of Galicia taken by S/Lt Léon Boittiaux during the winter of 1916–1917. (Coll Albin Denis)

Above: Russia was so short of airplanes that any captured enemy aircraft was immediately repaired and put back into service against its former owners, such as this Rumpler C.I in service during the winter of 1916–1917 with its new Russian roundels. (Coll Albin Denis)

Above: First contact with Russian soil for the N 581 and SOP 582 squadrons which landed in Lapland in April 1917.

2.3 – N 581 and SOP 582 squadrons

2.3.1 – Two squadrons formed as reinforcements in 1917

The Stavka was fully aware of the interest of the presence of French airmen in the ranks of its units and on October 28, 1916, asked that others be sent to it. On November 7, 1916, Count Ignatieff, the Russian military attaché in France, requested authorization to take advantage of the winter to send 40 Russian pilots, 10 observers, and 5 instructors to France to be trained in French schools and learn combat techniques. He was satisfied and the experience became widespread because a total of 111 Russian pilots were actually trained in French schools and received their military pilot's license from that date until the beginning of 1918.

As for the request to send additional airmen, it was carefully examined by the French general staff, who considered, after the advice of General Janin, that it was out of the question to send French soldiers to fight while leaving them dependent on the increasingly deficient Russian logistics. It was therefore decided to send two purely French squadrons which would leave for Russia with their own aircraft and their own air fleet, in order to be sure that they would have the means to fight.

The two squadrons were formed in Lyon at the depot of the 2nd aviation group, under the direction of Major Berger, who had returned from Russia. The first unit was the N 581 fighter squadron, equipped with Nieuport 17s and SPAD VIIs, commanded by Lt Louis de Gueydon and including among its 10 pilots S/Lt Georges Lachmann, a professional military pilot who had earned his pilot's license before the war and had won two victories at Verdun. The second was SOP 582, an artillery adjustment unit commanded by Captain Jules Balavoine, equipped with French-made Sopwith 1A2 two-seaters and comprising 11 pilots. Between the pilots, observers, mechanics, and general service men, the French detachment that left Lyon in March 1917 comprised 42 officers, 25 non-commissioned officers and 200 troops of all specialties, who were given a khaki uniform by the French intendance, and who embarked from Boulogne via Folkestone and Liverpool for Arkhangelsk in March 1917. Lieutenant Mayeur, head of the fleet of the two squadrons, testified about the trip:

"*We left Lyon for Liverpool where we boarded a Russian cargo ship [The* Divinsk, *Ed. note] loaded with ammunition, in which we stalled the forty or so aircraft and the 300 or so French servicemen. We left one morning escorted by British torpedo boats. We stopped in Scottish ports because the pressure of the German submarines was very strong and there was a constant risk of being torpedoed. Then we arrived in the north of Norway, the sea was covered with ice. We arrived in the Bay of Kola and then in Murmansk. The joy of the Russian peasants was intense when they saw French officers disembarking, they were absolutely flabbergasted and excited. They absolutely wanted to kiss us... And obviously these peasants with ice in their beards and the Russian habit of kissing each other on the mouth were not exactly to our taste! (...) During all this trip, the commander Berger had insisted that his officers learn a little Russian and had hired a Russian teacher from the Berlitz school to give us lessons. We were obliged to attend his lessons one hour a day, so that when we arrived in Russia we were able to express ourselves in Russian in a niggardly way. But it was very useful!*

Above: During this time, on the front, the airmen of the 27th KAO took an oath to the new Russian constitution adopted after the Tsar's ouster. The aircraft are Farman F.40 equipped with skis (Coll. Albin Denis)

So we arrived in Murmansk, a snow-covered port, but where a branch of the Trans-Siberian Railway ran to Petersburg, and from Petersburg to Moscow or Kiev. It took us about eight days to make the trip to Petersburg. From there we stopped in different places where we were received with extraordinary enthusiasm. The military bands played The Marseillaise *for us and the people said to us: you French, you made us see how to make a revolution, and we did it!"*

The political situation in Russia had indeed changed when the French detachment disembarked in Murmansk, since Tsar Nicholas II had been forced to abdicate on March 15, 1917. The vast Broussilov offensive, interrupted by the winter at the end of 1916, had certainly allowed the Austro-Hungarian armies to be pushed back a hundred kilometers to the west, into the Carpathian mountains, without regaining the ground gained during the 1914 offensives. But this leap towards the west of the Russian army was a Pyrrhic victory, obtained at the price of heavy losses because if 1 million Austro-Hungarian soldiers were put out of action or captured, as well as 300 000 Germans, 500 000 Russians were killed and almost as many deserted. This great battle was the one too many, which completed the disorganization of Russian society by increasing the misery in the cities and countryside of the empire during the winter of 1916-1917. A strike launched by the workers of the Putilov arms factory in Petrograd on March 3, 1917 degenerated and spread, leading to a general strike in the capital that the tsarist regime tried to suppress by force but without success because the local garrison sided with the insurgents. The strike spread to Moscow and a provisional government was formed from liberal deputies in the Duma. Pressed by his chief of staff, General Alekseyev, who urged him to abdicate in order to "save the independence of the country and ensure the safeguarding of the dynasty", Tsar Nicholas II was forced to relinquish power in favor of a provisional government headed by Prince Georgy Lvov, but within which the lawyer Alexander Kerensky would assert himself.

This new government aroused great hope in Russia, but its power was limited because it had to deal with local assemblies, the soviets, which were set up in the four corners of the empire and did not hesitate to challenge its decisions. The most important of the soviets, that of Petrograd, is infiltrated by the Bolsheviks who were waiting for their time...

The provisional government was determined to keep Russia in the war against the central empires, which would eventually cause its loss. For the time being, it was cleaning up the staffs where the time of the aristocrats was over. The head of the air force, Grand Duke Alexander, was dismissed on April 3, 1917, and after a three-month interim was replaced on June 22 by Vyacheslav Matveevich Tkachev, a pilot from the front who had won an aerial victory and was promoted to the rank of polkovnik. The head of the heavy air force, General Shidlovskiy, was dismissed on 18 May and heads also fell at the Stavka as the Tsar's former chief of staff, General Alekseyev, was replaced by General Broussilov on 16 April 1917.

As Major Berger had left for France to train the two squadrons he was to directly command at the front, he had to leave the command of the aeronautical mission to Major Alfred Bordages, with General Niessel also replacing General Janin as the Russian interlocutor at the Stavka. Meanwhile, the war continued for the pilots and observers already in place in the Russian squadrons and one of them, S/Lt Raymond Berné (2nd AAO) died on the field of honor on March 2, 1917. Flying as an observer on the Voisin-Lebedev n°422, he was escorting another Voisin on a photo mission but his aircraft was attacked and shot down by a German two-seater. Its Russian pilot, Praporshchik Grechkin, managed to bring it into the first Russian lines where it crashed. Russian soldiers rushed to the scene in spite of enemy fire and relieved Grechkin, wounded but not Berné, who was shot in the heart. They brought back his body in spite of violent artillery fire that destroyed the aircraft. Another French pilot, S/Lt Louis de Lignac, died in hospital on 7 July 1917 after being hit by flak on 22 June. Four pilots and three observers of the first mission were killed or captured in Russia, a loss rate of 35%, much higher than that of the squadrons of the French front... On the other

Above: Stationed in the city of Kiev before being sent to the front, the French airmen of N 581 and SOP 582 squadrons as well as their air fleet were visited by the French Minister of Armaments, Albert Thomas, on April 24, 1917. (Coll. Albin Denis)

hand, there are two aerial victories recognized by the Russian General Staff, and confirmed by the French authorities, to S/Lt Marcel Paillette who aboard his Nieuport shot down two enemy planes above Volodymyr-Volynskyï (West of Lutsk), during an escort mission, on March 22, 1917. Some survivors of the mission will later join the two squadrons, Sgt George Sausse (mechanic) at N 581 and S/Lt Poiré at SOP 582 – the latter having been dismissed from his unit by Grand Duke Alexander for having put in his place a Russian officer who pretended to teach him his job as an aviator...

At the end of April 1917, the train carrying the two French squadrons made its way to Ukraine, crossing the Russian countryside, far from suspecting that the country was crossed by a gigantic revolt where the peasantry appropriated the land of the aristocrats without the provisional government being able to do much about it. Lieutenant Mayeur, head of the park, testifies: "*We then arrived near Kiev where the Stavka, the Russian general staff, was located. We disembarked and prepared our equipment. The Russian offensive was scheduled for July 1917. I had foreseen the difficulties of transportation and I had asked the Russian General Staff for a train consisting of a sleeping car, and about twenty freight cars to make workshops of all kinds. I also had two wagons of food to feed all the men.*

Upon arrival in Kiev, one of the fighter pilots, S/Lt Marcel Bloch (ace with 5 victories against Drachens) was subtracted from the fighter squadron's strength to be assigned as a receiver pilot at the Anatra factory in Odessa, a factory run by the French pilot and engineer of arts and crafts Henri Jouffret. On his first flight, May 8, 1917, Bloch fell victim to the most serious danger of pilots on the Russian front, namely the deficiency of his own aircraft, and crashed at the controls of his Anatra DS Anasal on the edge of the factory grounds, as Marc Bonnier had done nearly a year earlier. He survived the crash but was seriously injured and had to undergo a trepanning operation which put an end to his career as a pilot. Two other French pilots who worked at the Odessa factory lost their lives a few months later: Jean Robinet fell on July 23, 1917 with his mechanic Ivan Omelin while delivering a two-seater Morane type P to the front, and then the pilot Lucien Chandon died in an accident near Odessa on November 26, 1917.

2.3.2 – In preparation for the Kerensky offensive

After the arrival of the French train at Kiev station, the French settled on the city's racecourse where they set up their air depot, waiting for the bulk of their equipment to follow behind them. The city was swarming with deserting Russian soldiers, as well as Austrian prisoners of war who went to their work freely... The French attended the May Day demonstrations held on their racecourse. S/Lt Louis Sers, park officer and assistant to Lt Mayeur, testified: "*From eight o'clock in the morning, huge columns of several thousand soldiers, workers, and women marched in front of the racecourse, then massed around the grandstands where speakers harangued them, enumerating the benefits they had the right to expect from the new government. The orator was listened to in silence; once the speech was over, the column reformed in order, without saying a word, and then circulated in the streets of the city preceded by a*

Above: The French convoy carrying the two escadrilles enters the Jeziezany train station on June 1917.

music that played an eternal Marseillaise, *definitively adopted as a revolutionary song. I have seen nothing more impressive than these silent demonstrations: one has the sensation of a force so enormous that once unleashed nothing could contain it. On this human sea it was still calm; but in the distance the storm was rumbling. One of our interpreters had the opportunity to question several demonstrators; he invariably got this answer: "Now that I am free, I want to enjoy life, the goods of the rich, the sharing of the land, and I don't see the need to return to the front." It is clear that men for whom neither patriotism nor a sense of military honor existed, could not reason otherwise. But the leaders to whom these words were reported were beginning to worry: the expected effect was far exceeded!*

If the French hear from several Russian interlocutors their

Above: Nieuport 17s of the N 581 squadron guarded by a sentry at Kolotche (today Holhocha, Ukraine) where the unit settled on June 20, 1917, just before the Kerenski offensive (Coll. Albin Denis)

Left: While the N 581 squadron was stationed at Holhocha, the SOP 582 reconnaissance unit settled not far away, a little further south, at Cheremkouf. Since it could not receive its aircraft, it lent its observers to a British squadron equipped with BE2s, such as this one. (Coll Albin Denis)

Below: Poor quality photo of one of the Sopwith 2A2 of the SOP 582 squadron (Coll. Albin Denis)

willingness to continue the war, they have many doubts about the effectiveness of the Russian army whose cohesion is only relative, and that the French socialist deputy Marcel Cachin (and future member of the French Communist Party) himself described as a "horde."

On May 24, 1917, the French received a visit from the French Minister of Armaments, Mr. Albert Thomas (a reformist socialist), who noted in his diary: "*We are going to the Berger group, which is well installed at the Kiev racetrack. I address a few words to the officers and soldiers. All seem to be in good moral condition. Some complaints for the soldiers (...) The installation is quite suitable. The minister and the French officers then went to lunch at the home of a wealthy local Russian family, the Balachewskis, where "the Berger mission seemed to be at home. In the afternoon, the minister and the airmen and soldiers of the Berger mission attended a public meeting in front of all the constituted bodies of the city".* He noted: *"An orchestra played* the Marseillaise. *It was a festive meeting. But our local Bolsheviks were still there and despite the kindness and eagerness of the first speeches, we were obliged to respond clearly. Our air force officers are very happy about this. Also the French soldiers, including some very satisfied and very happy Cegetists* [French socialist trade union, author's note]. *I do not know to what extent the Russian crowd is as well, but it is important to say what had to be said."*

With all their equipment having arrived as best they could in Kiev, the two squadrons then left for the front on June 11, 1917 by train, which was in fact the only reliable means of transportation in Russia, as S/Lt Sers, the park officer, who committed the imprudence of organizing a convoy on the road, pointed out: *"Roads did not exist in Russia; we had to*

Right: Rasputitsa ! Spring and autumn are the seasons of the year when travel on unpaved roads becomes difficult, owing to muddy conditions from rain or melting snow. The French mission experienced it while settling in Ukraine.

Below: This Nieuport 23 (n°2582) from escadrille N 581 was damaged on landing on June 9, 1917, during the Kerensky offensive. His pilot, Alberto Paolo Bucciali, was wounded. (Coll Albin Denis)

learn this the hard way, because no sooner had we traveled some one hundred kilometers on a gravel road, which was only remotely reminiscent of those in France, than we found ourselves on a dirt track. In summer, this track is sometimes practicable, but at this time, still too close to the melting of snow, it resembles much more a plowed field than a road. After some mireings from which we pulled ourselves only with great difficulty, force was us, having found horses, to make us drive back to our point of departure. We can see by this simple episode what a tour de force it is to move an army forward or backward in Russia, and what endurance it took for the soldiers at the beginning to wait for the arrival of supplies which were always problematic. Our journey, thus postponed, had to be made by rail, that is, it took us five days instead of one."

The French gathered in Boutchatch, where the headquarters of the 7th Russian Army, the spearhead of the offensive organized by the Stavka for the month of July. The park of the two squadrons then settled a little further west, in the town of Jezerzany.

Second Lieutenant Georges Lachmann, pilot of N 581, testified: "*We arrived at the beginning of June in Boutchach (Galicia), a little behind the front of the 7th Army, the place chosen for our park-train. At that time the Boche were quite fierce and came every morning to carry out bombardments at low altitude, causing enormous devastation (This village of Boutchach*

Above: View of Georges Lachmann's SPAD VII on June 26, 1917, or what is left of it, after it crashed into some trees, after having flamed a Russian sausage...! (Coll Jack Herris)

was partially destroyed by aerial bombardments). We felt surrounded by Kraut spies, who swarmed all along this front and we expected to be bombed the very morning our trains arrived. To avoid it, I had two Nieuport 110 hp assembled and adjusted at night (a real tour de force when you know the condition of the planes when they arrived in Russia, after four or six months of travel) and in the morning we were, a comrade and I, on the lookout for the first of these gentlemen: nobody came."

With the arrival of the equipment that the soldiers of the park bring up to Jezerzany, the two French squadrons are formed and the N 581 fighter settles on June 20 15 km north in the village of Kolotche (currently Holotcha, sometimes spelled Xolxotche in French documents), while SOP 582 settles a little less far in the neighboring village of Cheremkouf (Czeremchom, currently Shveikiv), as evidenced by S/Lt Louis Sers: "*The corps squadron was installed in an Austrian castle, in a charming site. Captain Balavoine, who commanded it, had to provide photographic and topographic services, adjustments, and reconnaissance with four cameras (the others were still in Arkhangelsk). There was no lack of work for him, as the front held by the army was more than 40 versts long. Less well shared was the fighter squadron, commanded by de Gueydon; it had, I do not know why, been placed on a hillock on all sides in the exposed sun, without a tree, without a house, without a drop of water.*"

The first war flights then took place - SOP 582 took a head start by lending its pilots to a Royal Flying Corps reconnaissance squadron flying BE 2-seaters. On June 18, 1917, Lieutenant Maurice Denis, an observer, flew a mission on a plane piloted by British Lt Gordon Smith and was attacked by two enemy planes; with the fire of his rear machine gun he shot down one of them which fell into the enemy lines over the Michisuv train station and received a citation in the army order for this exploit Then, on June 28, 1917, S/Lt Pierre Dufayet, probably flying a Sopwith with an unknown observer, shot down an enemy aircraft over Gusyatin according to the Russian communiqué, but did not receive any French citation.

The French fighters of N 581 were not left out to carry out flights but remained empty-handed... They flew Nieuports and had for the moment only one SPAD VII reserved for the best pilot, S/Lt Georges Lachmann, who testified: "For a few days, we saw quite a number of Austrian aircraft, which were not very bright in appearance, but very calm and prudent; we could not approach any of them at that time without having a jam after two or three shots. The enemy airfield was about twenty kilometers to the north and we were going to wait for our partners over their own ground; very few dared to rise in our presence and all those who were in the air at the time of the flights returned to the nest by crawling.

The French Nieuport 17s were equipped with Le Prieur rockets designed to shoot down German captive balloons and went on the attack to blind the enemy, in the imminence of the Russian offensive. But without success... Lachmann wanted results and decided to act.

"*At that time the Russians were preparing their great spring offensive and were counting heavily on our presence for all air services. Something had to be done; our unfortunate fights had not produced any results; our pilots, Second Lieutenants Coudouret and Jalaguier, had missed drachen attacks. On June 25, I saw Major Berger and decided with him to carry out a drachen attack on the ground. So I prepared myself!*

On June 26, 1917, at two o'clock in the morning. Departure before daylight. Lieutenant de Gueydon was to accompany me to avoid a too unpleasant encounter. The weather was gray, foggy, with low clouds and very little sunshine. A few days before, I had spotted a drachen in a small patch of fir trees a little south of a large lake. From the start, it was impossible to find my support plane, which had gone I don't know where: I decided to leave alone and to avoid any enemy marauders in search of good fortune, I climbed above the clouds.

On the ground, in our rear-front, the few lights that were shining seemed to me as many flashing fireflies. Magnificent spectacle! Which, found in all my morning flights, never wearied me! Under me, the big white flakes passed with speed under the wings of my SPAD which walked towards its destiny! He didn't think he was so close to his end!

So I walk, helping me to direct myself with general points that I knew well; after some time, a lake, the one I was looking for, appears to me through a small clearing! A little push on the left rudder, a little rudder movement to the left, a few swings to the right, to the left, a clearing. Here is my prey; there, in the middle of its small belt of large fir trees, my drachen, my Boche sausage! I feel it to me, nothing can save it! My altimeter? 1400; a glance around, everything is calm, let's go.

Full engine my SPAD dives, dives towards its destiny. Here I am at a good range, at 20 meters from the ground. I press my contact, my rockets leave and all enter in the balloon. It flames

Right: View of a captive balloon used by the Russian army. We note that it is a model H, a French copy of the German Drachen, which is no longer in service in the French army since 1916, replaced by the Caquot balloon, which is much more efficient in its resistance to the wind (Coll. Jack Herris)

like a torch. Immediately I straighten up and, in the same second, what am I saying? In a thousandth of a second, I see in front of me a hill with trees. With my immense speed acquired during this long dive, I enter the trees! I feel my camera catching and start to turn left! It's the big log! I cut my engine not to catch fire. Immediately, I feel myself thrown in the air like a ball at more than 30 meters and I find myself, dumbfounded, face down.

Completely covered with earth, not being able to stand up, having pain everywhere, I put myself on all fours and cleaned my eyes one after the other; nothing, I see clearly. I bled all over my face. My chest hurts, my back hurts! My left hand hurts. I take off my glove, my index finger is broken, completely turned over; I put it back in place.

Then, all of a sudden, I think, and a frightful fear comes to me: I burned the Boche ball, my apparatus is broken. So I am a prisoner? I stay for a moment absolutely stunned by this idea, then suddenly I turn around, I see in the middle of the trees the balloon burning, 30 meters away from me my camera reduced to the size of a small trunk. Unable to walk, I drag myself to hide. I hear shouting and running behind me, I am done for! Nothing to do! I lay down on my back and waited. I am surrounded. What do I see? Russian caps! In German they ask me: "Do you speak German?" I answer, "Yes, but not much." Immediately I hear in Russian: "He is a German, he is a German." In Russian I reply: "I'm not German, I'm French, but why do you have the Russian uniform? You are Germans, the balloon I just burned is German. - No, we are Russians and so is the balloon."

Astonishment! I had just burned a Russian sausage! What a story! Especially at that moment! Up above the clouds, at night, with mist on the ground, I must have mistaken one lake for another and gotten lost. This is the result.

They put me on a stretcher and we went up and down through ravines. What I take! Finally, they put me in a cart without springs, pulled by a horse (you have to know the roads). During all this time, I think: I have just burned a Russian balloon! I have just burned a Russian balloon! A passing ambulance driver bandages my head for the first time, then we go on our way. More than an hour and a half later, we arrived at an ambulance where a Russian sister undressed me, cleaned me and bandaged me. I hurt all over.

The doctor arrived; I had a broken left index finger, a cut above my upper lip, one near my left temple, one on my chin and one on my neck. I pretend that I have nothing in my chest, nor in my kidneys. My legs are blue, my face is blue-black, the whites of my eyes are bloodshot.

Soon the captain of the balloon came along and, believing me to be a Kraut and shooting at me with his revolver, put a bullet in his leg. We shake hands and laugh. Then it was the Russian commander of the artillery in the sector. He tells me that, despite the mistake, my attack was surprisingly brave: charming man! Finally, de Gueydon arrived a little distraught: not seeing me, he had landed; as I had not returned, he had left three times to look for me and everyone in the squadron thought I had fallen into the hands of the Boche.

Major Berger came to see me at the English hospital where I had been taken two days after the accident and gave me a copy of a letter addressed to him by the general commanding the Russian army. Here it is: "June 14, 1917 [June 27 according to the French calendar, Ed.] I note in the report I received from you that on June 13 [June 26] you gave Lieutenant Lachmann

Above: The beginning of the end... On July 21, 1917, the French squadrons receive the order to withdraw to Jezerzany, which is done by car, truck, and horse. (Coll. Albin Denis)

a very dangerous mission full of risks, namely: to destroy an enemy drachen even on land. The atmospheric circumstances caused an error of orientation, the result of which was the attack of a Russian drachen at very low altitude. I understand that Lieutenant Lachmann is very affected by this error. I beg you to convey to him all the esteem I have for his valor and bravery, to wish him a speedy recovery and to assure him that mistakes of the kind of the one of June 13 are always possible, not only in the air, but even on the ground where the conditions are, however, more favorable, and that they are excused by all those who fight. General Z..."

Lachmann's injuries, although slight considering the circumstances, were not negligible: he suffered violent contusions to the thorax, a dislocation of the coccyx, a fracture of the left index finger, a fracture of the upper molars, and a cut on the underside of the chin, which earned him 3 stitches. He never regained the full use of his left hand and was awarded a 10% disability pension as a result of this accident.

2.3.3 – The failure of the offensive led the squadrons to retreat four times

Lachmann returned to his squadron on July 12, 1917, imperfectly recovered, and took over as interim commander because Lt de Gueydon was on leave in Kiev, from which he never returned, leaving Lachmann in command for the entire campaign. The great offensive known as Kerenski began on July 1, 1917 in Galicia, with the 11th, 7th, and 8th Russian armies from north to south. The Russian troops initially advanced about 20 km into the enemy lines. But undermined by the Bolshevik propaganda and even more by all kinds of shortages, the Russian soldiers mutinied and the whole front collapsed from July 16, allowing the German and Austro-Hungarian troops to counterattack on July 18 and to make an advance of nearly 240 kilometers in Ukraine. This rout cost General Broussilov his job, and he was replaced at the head of the Stavka by his former deputy, General Lavr Kornilov.

On July 21, 1917, N 581 and SOP 582 were ordered to evacuate their airfields threatened by the enemy advance, and to go to the village of Jezerzany (today Ozeryany), to join their park. The retreat was carried out by air for the aircraft, and by road for the rest of the personnel, as Lachmann testified: "The entire automobile convoy left by road as well as our Russian convoy with horses and carriages. All this in the midst of the mad rush of the Russians, artillery, ambulances, trucks. Nothing can give you an idea. Suffering from his wounds, he underwent a grueling retreat, having to run at every opportunity to go to the aid of every trainard in his unit, aircraft or vehicle. The retreat was carried out in good order for the N 581, but a little less so for SOP 582, which was obliged to leave behind two of its aircraft (n°49 and 1517) which were conscientiously demolished so as not to be left to the enemy.

The field of Jezerzany is also the one of the 1st combat group of the Russian fighter (BAG, Boevaya Aviatsionnaya Gruppa) that the ace of the aces of the Russian fighter, Major Alexandr Kazakov, then holder of 10 aerial victories, leads. He made a strong impression on Lachmann who wrote: "We were invited by his officers, as French comrades-in-arms. There I saw him. I always remember this first vision: tall, blond, oval face with prominent cheekbones, long moustaches that he always raised with a nervous movement.

Above: A French Nieuport in Jezerzany on June 21, 1917.

Above: The field of Jezerzany is also occupied by the 1st Combat Group of the Russian Air Force (1st BAG), gathering three fighter squadrons, and led by Captain Alexander Kazakov. While the staff aircraft (the four planes in the foreground) bear a black skull and crossbones on a white background, those of the 19th KAO bear the same skull and crossbones with white reversed colors on a black background. In the background, 6 Nieuports of the 4th KAO can be seen wearing a white six-pointed star on a red background, then, barely visible, the aircraft of the 2nd KAO wearing a yellow ace of spades on a black background, or vice versa. (Coll Jack Herris)

The characteristic of this face: large blue eyes that revealed in full the good Slavic soul, magnanimous in normal life, as much as ferocious in combat."

The passage to Jezerzany is only ephemeral because from the evening of July 22 the two French squadrons must go to Cherkoff (today Chortkiv) about thirty kilometers east, to settle July 23 in Gusiatin (Houssiatyn) 20 km further east, for only one week, because on July 30 the French group settled more permanently in Kamnicz-Podolsk (Kamianets-Podilskyï), a city built around a fortress near the northern tip of present-day Moldavia, on the northern bank of the Dniestr River. They were then placed at the disposal of the 3rd Russian army corps. It was in this city that the main echelon of their fleet, which had undergone a thousand adventures, joined them. When the order to retreat came on July 21, Lt Mayeur, who remained in Jezerzany to collect the planes and equipment of the two squadrons, evacuated all the equipment of his fleet by train to the east. He entrusted it to S/Lt Louis Sers, who was directed by the Russian authorities in Boutchatch, 70 km to the south, to the town of Czernowitz. There he discovered a city in chaos, crowded with refugees, where he had to camp in his wagons loaded with equipment. He only managed to leave by forcing his way onto a locomotive and setting up his machine guns against the local stationmaster. He then managed with great difficulty to make his way through the Russian railway network, which was saturated with refugee convoys, and reached the Kamnicz-Podolsk airfield, where he fortuitously learned that the French squadrons were located. The air force, finally reunited, was ordered a few days later to take shelter in the town of Proskourov (today Khmelnytskyï), located 80 km to the north.

On August 2, 1917, shortly after the French squadrons became operational again at Kamnicz-Podolsk, Lt Maurice

Above: In parallel to the action of the N 581 and SOP 582 squadrons, the war of the 1916 Berger mission continued... And so do their losses. Here, a Morane LA crushed on July 23, 1917 by pilot Jean Robinet, from the Odessa factory (Coll. Albin Denis)

Above: On July 22, 1917, the French squadrons and their fleets left the city of Jezerzany by train for a long journey across Russian train lines, in which two echelons of their Air Park had many adventures.

Denis, a pilot with SOP 582, distinguished himself once again by shooting down an enemy aircraft near the village of Skala. During this time, Lachmann took advantage of the situation and went to Kiev to be treated in the city's hospital, and to take possession of a new SPAD VII. He tried it out over the city of Kiev on August 5, 1917 and decided to fly for his squadron the next day. The trip had all the adventures of Michel Strogoff: the radiator of the aircraft malfunctioned and he was forced to land seven times in the wilderness, either to refuel or to make makeshift repairs. During a forced stopover in a small Ukrainian village, the peasants refuse to let him leave, taking him for a German... After several telegrams to clear up the misunderstanding and 24 hours of immobilization, he was able to leave and arrived after several other adventures on August 8, 1917 in Kamnicz-Podolsk, in a very bad mood if we are to believe his correspondence: "I still had to take revenge on the Krauts for the unfortunate mistake of which I had been a victim. That was not going to be long!

Left: The men-in-shadows who allowed the French units to remain operational during all the campaign: lieutenant Georges Mayeur (2nd from right), chef of the French air park, and his second sous-lieutenant Louis Sers (extreme left). These man ruled the armored trains that contained all the supplies of the units. The picture was taken in August 1917 in Proskurov (today Khmelnitsky, Ukraine).

Above: The 581 squadron failed on August 12, 1917 in the city of Kamienets-Podolsk, from where the first aerial victories were won by the Lachmann/Coudouret duo (Coll. Albin Denis)

2.3.4 – French fighters win 8 air victories

The French fighters increased the number of flights on the new front line that was established along the Zroutch river, a tributary of the Dniestr. The sky was quite empty compared to the French front... Lachmann, by dint of obstinacy, found something to open his score: "On September 1, I went to the lines in search of something. I saw a drachen at Melnetza, 12 kilometers inside the lines. I returned to the field to take rockets and set off again: I massacred it at close range, only one rocket leaving out of the 6. In all, 20 minutes from departure to landing! This time, the Drachen belongs to the enemy... And more precisely to the Austrians: the Kuk Ballonkompagnie 22 loses one of its balloons, its observers, the LdR Egon Werthheimer and Fuerwerker Frantz Antos parachuting unhurt. This balloon is not mentioned in an official communiqué of the Russian troops, but will be recognized by a mention in a later Russian communiqué.

If the French fighters had the opportunity to celebrate their first official success at the squadron's bar, the news coming from Petrograd was not very pleasing. Wanting to subdue the Petrograd soviet which had launched a press campaign against him, General Kornilov marched troops into the city contrary to the orders of the Kerensky government and initiated a military coup. The affair failed and Kornilov was deposed, but it weakened the Kerensky government, which had to call on the Bolsheviks, who were now armed and ready to act... Their leader, Lenin, came out of hiding. On the front, the little discipline that remained in the Russian armies and that Kornilov had tried to restore disappeared with him. Kerensky appointed General Nikolai Dukhonin as head of the Stavka, the fourth man to succeed him in this position since the beginning of 1917...

This did not deter the French aviators from flying as much as they could. Lachmann had another success: "September 15, sport! I went out and, once I was on the lines, I heard Russian gunfire in the sector of the 8th Army, the one next to us south of the Dniester. I ran there and found a beautiful yellow biplane with nice black crosses. In two steps I am with my SPAD under his tail without him having seen me and I send him at least 40 shots. He starts to turn and the observer shoots me in his turn. A few somersaults to get out of the way, I replace myself and withdraw: he starts to spiral down and I follow him like a leech, spraying him every time possible at 250 meters. After a small syringe, I see a grayish mass falling out of the aircraft while spinning, having the shape of a cross; I think for a moment of a parachute, but no, it can only be the observer who, killed or wounded during a sudden turn, is

Left: This photo, of poor quality because it comes from the magazine "*La guerre aérienne illustrée*", shows the SPAD VII of S/Lt Georges Lachmann landing on August 6, 1917 in Vinitsia (Ukraine) on his way back from the city of Kiev, a step among others in a hectic journey back to his squadron in Kamienets-Podolsk which he finally reached on August 8. His aircraft is decorated with his personal symbol, the question mark, and also carries an additional Lewis machine gun mount on the upper wing, which is a local DIY job as SPADs were not equipped with them. (Coll Albin Denis)

Above: The SPAD VII of Second Lieutenant Louis Coudouret, probably at Kamienetz-Podolsk between August 12 and November 15, 1917. (Coll Albin Denis)

emptied of the aircraft. The pilot launches some rockets to call his comrades. Finally, after 20 minutes of fighting, I shoot him down in his front lines (We found the rear machine gun and not the observer.)" The official communiqué of the 7th Russian Army Aviation, dated September 17, 1917 states that "During the week, an enemy aircraft was shot down by Lt. Lachmann in the Cherovtsé (Shyrivtsi) area, according to the observation of the troops, this plane fell in the front lines, at the village of Chilovsté (Klishkivtsi). This plane dropped a machine gun in the village Ornichard." There is no record of a downed aircraft in the German or Austrian archives, but it is officially recognized and earns him a citation in the army order dated October 30, 1918.

On September 20, 1917, it was S/Lt Paul Jalaguier's turn to shoot down an enemy aircraft at Chabarivka, which was recognized by the communiqué of the 7th Russian Army but did not earn him any French citation. Lachmann obtained his 3rd victory on the Russian front on October 3, 1917, which made him an ace because they were added to the 2 victories he had already obtained at Verdun: "New piece, after six hours of flight during the day. The Russian 7th Army communiqué dated October 4, 1917 states, "4 encounters in the air, during one of them, the valiant French pilot, Lt. Lachmann, shot down his second enemy plane, which fell

Above: Another SPAD VII of Second Lieutenant Louis Coudouret, touching with his hand the propeller of the aircraft, which was also equipped with tubes for Le Prieur rockets. (Coll Albin Denis)

Right: View of a SPAD VII (n°338) of the N 581 squadron probably at Kamienetz-Podolsk, in autumn 1917. As is often the case with French aircraft in Russia, it bears no distinguishing marks. (Coll Albin Denis)

in the area of the village of Malinsté (Malyntsi), behind the enemy trench lines." German records tell us that a two-seater from Flieger Abteilung 242w was indeed damaged that day.

On October 8, 1917, Lachmann and S/Lt Coudouret, flying in concert, scored another victory against "a two-seater that crashed on fire just in front of the Russian trenches." The 3rd Caucasus Corps communiqué issued the same day stated, "An enemy plane was attacked by our fighter, after a short fight in the air, the enemy plane fell at the northern end of the village of Zaliesce (Zalessky, near the Zbrousch River), and caught fire."

On October 11, Coudouret narrowly missed being shot down when he was intercepted alone by three Albatros D.V. single-seaters, initiating a spinning fight between fighters that was quite rare on the Russian front: "I have, against three of these aircraft, the hardest fight of my aerial career! It lasted 45 minutes! Spins and slides are useless. These gentlemen accompany me and strafe me up to 40 kilometers in the lines and 500 meters high. My SPAD, once again, was a skimmer: candles crossed, cables cut, engine riddled, incendiary fuses torn off, uprights cut. There are days when one is astonished to find oneself on the pasture without any pain!

Then, on October 16, 1917, Lachmann renewed his success during a glorious day that he initiated at 4:00 pm by shooting a two-seater that he had to abandon in enemy territory, at Yagelnitsa. The victory was not confirmed, but the Austrian archives tell us that an Albatros C.I (registration number 169.09) belonging to Flik 11, landed in its lines with its observer killed (Leutnant in der reserve Ludwig Theimer), while its pilot, Feldwebel Rudolf Mueller, was unharmed. Later that afternoon, at 6:20 p.m., he spotted a drachen in the air at Malyntsi, not far from where he had shot down one on September 3, and immediately went on the attack: "I shot

Above: Lineup of Sopwith 1A2 of SOP 582 squadron probably Kamienetz-Podolsk, autumn 1917. (Coll Albin Denis)

down a new Boche drachen at a height of 200 meters, with a single rocket." The Austro-Hungarian archives do indeed tell us about the loss of the C 24 balloon... belonging, as on September 3, to KuK Ballon Kompagnie 22, and mounted by the same two observers, the Leutnant in der Reserve Egon Wertheimer and the Feuerwerker Franz Antos, whom Lachmann once again forced into a parachute jump! The communiqué of the Chief of Artillery Reconnaissance of the 12th Russian Corps of October 18 states that "*The French pilot Lieutenant Lachmann had a fight with two enemy planes and pursued them far into the rear of the enemy territory; these planes landed in Juaguelnitza. That same day, Lieutenant Lachmann set fire to his second enemy drachen in the Melnitza sector, where the two observers were burned. This is the fifth victory in Russia for Lieutenant Lachmann, who shot down three enemy aircraft and set two drachens on fire.*" On October 27, 1917, he nearly scored another success by attacking a drachen at Cernovitz. The observer jumped by parachute while Lachmann machine-gunned the envelope of the balloon... Which did not catch fire, and for good reason: his mechanic did not put incendiary bullets in the ammunition of his machine gun!

During the night of November 6 to 7, 1917, an event takes place in Petrograd that will change the face of the world, namely the insurrection launched by the Bolsheviks who seize power by force, at least in the capital as well as in Moscow, because many local soviets did not recognize Lenin's new government. Lenin ordered General Nikolai Dukhonin to immediately start peace talks with Germany: he refused and was dismissed, being replaced by the Aspirant Nikolai Krylenko, a Bolshevik militant of the first hour. But before Krylenko arrived at the Stavka in Mogilev, Dukhonin, who remained in his post as a transitional measure, had General Kornilov and his supporters (including General Denekin) released from prison: they formed the first White Army in the Don.

With the political situation uncertain, the two French squadrons were ordered to withdraw to Volochyst (Volochysk) on November 15, 1917, which was to be their last battleground. The city is about 80 km north and 40 km west of Proscurov where their park is located. In Volochyst they discovered not far from their field the armored car company of the Belgian army fighting on the Russian front. The French pilots kept their posts while waiting for the evolution of the events. Coudouret notes: "*The days passed and the rain, obstinate, fell without stopping. We were gnawing at our brakes. Finally, on November 23, 1917, a little before noon, a tear widened in the gray veil of clouds and the sun came out! We immediately abandoned the bridge we had started. As I sit in my cabin, idling my engine, a few white flakes of 75's bloom above Volochyst, high in the sky! Already the mechanic, removing the chocks, showed them to me: - A Boche! What luck! I take off and, the eyes fixed on the small black point which, flees towards the lines, I pull vigorously on the stick of my SPAD. "Will I reach him in time? Suddenly, the Boche - I am now a few hundred meters above him - turns around and heads back to Volochyst. I let him get stuck, while placing myself between the sun and him. He doesn't seem to have seen me and quietly goes his way. It would be really funny if I shot this Kraut above our field! Here it is! He is at 4.800 meters. I am 200 meters above. I dive, reach him with a few wingbeats. At my second salvo, I see the gunner abandon his weapon. Several of my bullets hit the turret and the machine gun pivot. However, the pilot seems determined to continue the fight, and, turning suddenly, tries to take me under fire with his front machine gun! Here I am obliged to send him again - as a persuader - some bursts. At 800 meters, I see the silhouette of the observer raising its arms on the luminous screen of the sky and making "kamarad! I get closer to him and ask him to land. But, far from obeying, the pilot pushes forward and tries to escape me in a sudden turn. With a last burst I punish him! The plane catches fire!*

The fight is over. The pilot lands in all haste. Blinded by the gasoline smoke and seriously burned, he does not see the ground, gets stuck in it and capsizes. Without delay, I landed in the same field and helped my unfortunate opponents to free themselves from the debris of their plane. They were not as badly hit as I had expected. At that moment, several hundred tovarichs rushed in from all sides, mistook me for a Kraut, rushed at me, fists raised, and seemed quite willing to lynch me! I want to explain myself! My bad Russian only makes the situation worse. Finally, a non-commissioned officer who speaks French makes his way to me. Everything is arranged, and these big boys who wanted, a moment before, to beat me up put me up on their strong arms and make me jump. It is for them, I know it, the greatest mark of honor and admiration, but I admit that the manner seems to me a little strong!

When after a few minutes, with broken kidneys, I finally manage to put my feet on the ground, I make my tovarichs understand, not without difficulty, that I do not want my prisoners to be harmed, and I immediately entrust them to two Russian non-commissioned officers to take them under escort to the nearest ambulance, and from there to our kitchen! They leave, heads down. But, an hour later, when they sit down at our table, face and arms bandaged, they have regained all their arrogance, especially the observer, a former infantryman, who boasts of his five wounds. He even says to me: If my machine gun had not been hit, I would have shot you like a rabbit. - It is very regrettable for you," I smile, "that it is the opposite that happened! - Oh! he resumed, "for whoever wants, the Russian lines are not impassable! In the evening, they leave under guard for the interior."

The German archives indeed inform us of the loss of an Albatros C.X of Flieger Abteilung 242 whose crew was captured. The 7th victory of the N 581 squadron was the 4th personal victory of S/Lt Louis Coudouret, who received a commendation in the Army Order on October 30, 1918.

Louis Coudouret went on to win the last aerial victory on Russian soil during the First World War: "Already, the rain began again, fine and persevering. Eight days went by without a break in the weather! Finally, on December 1, 1917, the eve of the armistice, a little before nightfall, we can take to the air, and I fall by surprise on an Albatros D V which was walking peacefully, at 2,400 meters above Volochyst, as a winner would do in a conquered country! In eight rounds, I shot it down in front of the first Russian trenches. This was to be our last battle!

December 2, 1917 is indeed the application of the general ceasefire unilaterally applied by the People's Commissar for War, Nikolai Krylenko, pending a peace treaty that leaves to negotiate Leon Trotsky. On December 3, 1917, General Dukhonin, who left the Stavka after Krylenko's arrival, was murdered by Bolshevik soldiers at the Mogliev railway station.

2.3.5 – Withdrawal from the front

The French witnessed the decay of the Russian army and saw the soldiers and officers remove their rank insignia and all their decorations following the Bolsheviks' instructions.

In this chaos, Lachmann, squadron leader (and promoted to the rank of lieutenant on December 24), organized the evacuation of the French detachment to the town of Loubny in order to put some distance between them and the German army: *"December 1917. The armistice has just been signed between the Russian provisional government and the Boche. Bolshevism is wreaking havoc in the still somewhat disciplined army corps. Guns, horses, etc., are sold at a low price or given away for a few bottles of alcohol; the greatest fraternity reigns between the former enemies. The Belgians evacuated a fortnight ago. We have to leave, otherwise we will be "picked up", one fine morning, by the Boche, in our beds! After innumerable steps, I was able to organize a train and take on board all our trucks, equipment, kitchens, and horses. (The squadron, with the Russian personnel assigned to it, amounted to more than 100 men). Passing through Proskourov, I added the park and the whole thing went to Kiev and arrived there in 4 days! There, the greatest effervescence reigns! The government of the Ukrainian republic was betraying in full view of everyone and preparing the coming of the Austro-Boches. Loubny, a small town located 140 versts east of Kiev, saw us arrive on December 25. On the 26th, a small revolution broke out among our Russian soldiers. Bolshevism was making its way! I kicked out 15 of them. No more stewardship, no more money or supplies to feed and clothe the rest; the sale of the few Russian horses was used to bring the ordinary cash register back in its expenses! One felt the storm in the air, everywhere! I was ready for new adventures!"*

An echelon of the park that Lt. Mayeur directed was not on the trip and escaped only after many adventures. "*I was with my train and couldn't do anything. We were threatened by the thousands of Austrian and Turkish prisoners who had been released and were prowling the countryside. Berger ordered the 2 squadrons to return to Kiev. I had a considerable amount of material in my camp, planes being repaired, vehicles being repaired, tents, food. And I had a hundred German and Turkish prisoners that I had asked the Russian commissariat to use as workers in the camp. I had to gather the prisoners and tell them: your friends are coming, you are free. The Germans did not made me repeat this twice, nor did the Austrians. There was an incident: I had, among others, a dozen Turkish prisoners with a sergeant. When the Russian administration gave them to me, they were unfortunate people who were starving, there were cases of scurvy. The Russian administration was not very generous with them, for example they gave me a sheep's head and a helmet of beans to feed 50 men, which was ridiculous. As I had plenty of food, I decided to feed the prisoners in the same way as the French mechanics. When I told the Turks "you are free", their sergeant came in front of me and bowed down. He said to me: "You have been good to us, we will not leave you. We know that you are in danger from the released prisoners, we will defend you. If you give us weapons, we will stand guard." I was a bit puzzled... On the other hand I had sent most of my men away and I only had about twenty Frenchmen left. So I was obliged to accept the offer of the Turkish sergeant. I made them give men and these unfortunate men stood guard 24 hours a day, keeping out the Austrians and*

Nieuport 17 n°2592 of the N 581 squadron based in Kolotche (currently Holocha, Ukraine) and flown by Warrant Officer Paul Jalaguier, mid-June 1917. The aircraft is equipped with Le Prieur fuses.

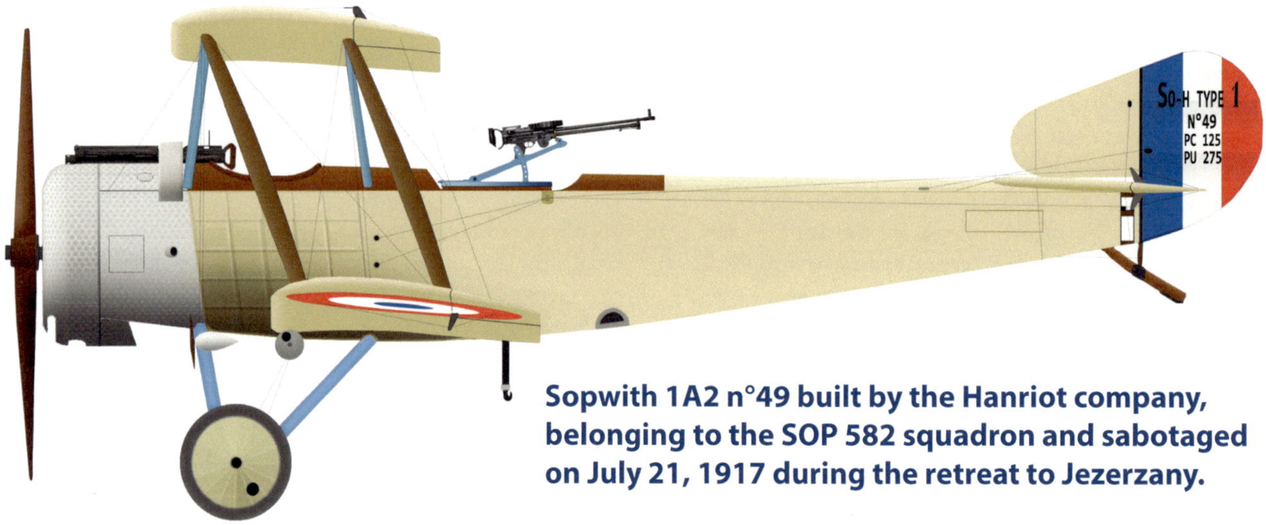

Sopwith 1A2 n°49 built by the Hanriot company, belonging to the SOP 582 squadron and sabotaged on July 21, 1917 during the retreat to Jezerzany.

SPAD VII n°338 of the N 581 squadron during the summer 1917.

SPAD VII of S/Lt Georges Lachmann decorated with his personal question mark. This is most likely the aircraft with which he won his first official victory on September 1, 1917, from the Kamianets-Podolsk airfield.

SPAD VII of S/Lt Coudouret, Kamianets-Podolsk, fall 1917.

Another SPAD VII attributed to S/Lt Coudouret.

Above: A French mechanic in Loubny on January 1918.

Germans who approached the camp!

They took the train away from me, I was lost with my 20 Turks in the Russian immensity. I had with me a man whose nationality I had never known, Romanian or Russian. But he was a guy who spoke all languages, resourceful. I had given him a French soldier's capote. One day he said to me: "Do you want to buy a train? There is one over there coming from Tarnopol. The mechanics would like to go home... I will bring them to you." Then I see my man arrive with the two mechanics. And then I bought the train for two liters of vodka and two boxes of 7.35 mm cartridges. And I got my train, but it was a question of how to make it work, because my French mechanics were aviation mechanics, not train mechanics. But fortunately I had two navy mechanics, who told me that we could work something out. So I had all my equipment loaded and we left without a map for the east. We were still puzzled at every switch. On the way we picked up some Russian officers abandoned by their troops. After about eight days we arrived in the very south of Russia, near the Romanian border. The name of the station escapes me... There was still some semblance of organization there. I was given real mechanics and a train. And after eight days we returned to Kiev."

In Loubny, 582 Squadron had practically no more planes and on December 18, 1917, General Niessel ordered them to leave for Simferopol, in the Crimea, to put their piloting skills at the service of the local aviation school. These pilots did not stay there for long, as events were accelerating. With the Bolshevik coup d'état, the local Ukrainian soviet, which calls itself the Rada (assembly), decides to secede from the Russian Empire by proclaiming the country's independence. It tried to recover military equipment and men to quickly create a national army, with the aim of resisting the Red Army which marched on Kiev... Because Lenin's government, which proclaims loudly the right of the people to self-determination, denies this right to the Ukrainians by specifying that the Russian and Ukrainian proletariats must remain united to fight against capitalism. He recruited an army among the Bolshevik sailors of Kronstadt and entrusts the command to Mikhail Artemevich Muraviev, a colonel of Petrograd gendarmerie. This Red Army arrived by train to the east of

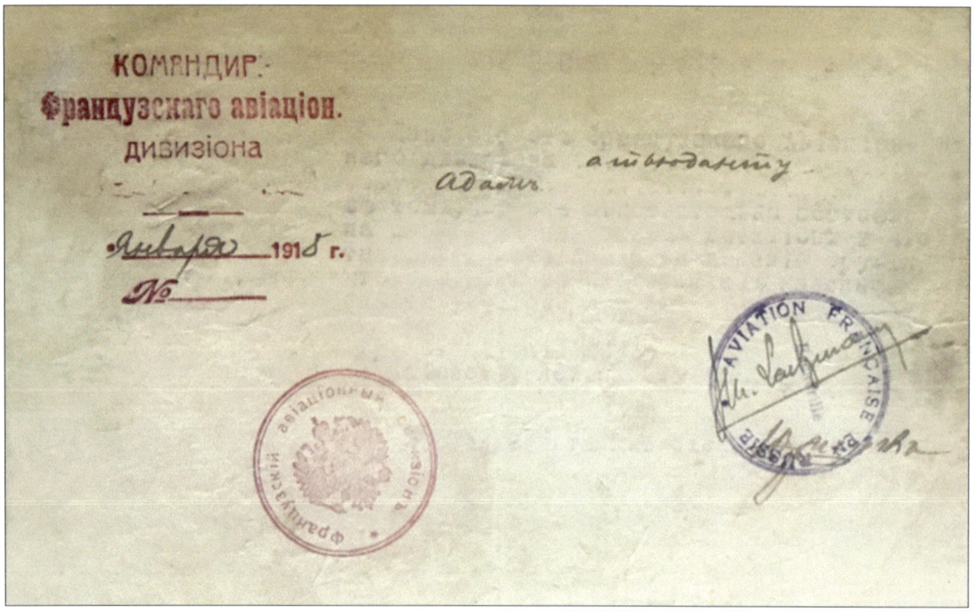

КОМАНДИР
Французскаго авіаціон.
дивизіона
Января 1918 г.
№
AVIATION FRANÇAISE

Left: Pass granted to the French by the Bolshevik forces, also signed by the head of N 581, S/Lt Georges Lachmann (Coll Matthieu Comas)

Ukraine and installed a provisional Ukrainian communist government in Kharkiv, then moved towards Kiev in armored trains. On January 25, the Red Army arrived in the town of Loubny, where the French airmen were located.

The Red Cossacks plundered the city without any restraint... S/Lt Sers testifies: "*As soon as they heard the rifle butts announcing the visit of these bandits, the inhabitants thought they were lost: there was no hiding place that they did not open for fear of offending such irritable characters. Not far from our park, there was a huge Red Cross depot; the Bolsheviks immediately requisitioned it, and then sent all the food supplies to Poltava. Soon, having learned that there were French troops there, the commander sent me a delegation to ask us what our intentions were: "We are still the allies of the Russians, and we do not concern ourselves with their internal politics," was the answer... And for once, they left us alone. But a few days later, my comrade Lachmann, whose squadron had also left the front and had retreated a few kilometers from us, sent one of his planes to Kiev to take orders, look for money, and inform our general that we could no longer think of joining him, since the way was taken by the Bolsheviks who were leaving to conquer Kiev. Informed of this step, Muraviev gave the order to arrest and disarm all the French.*

I will not describe here the heartache I felt when I saw this band of energetic people claiming the machine guns I had, as well as my weapons and those of my men. Fortunately, none of those who had come to get them spoke French; I therefore had time to make my men understand that there was nothing to be done against the number, that any resistance was useless, but that I intended that no weapon should be delivered in a state of service. Then I moved, followed by the Bolsheviks, towards the end of the train opposite the one that contained our armory. During this time, the big hammers flattened on the anvils the guns and the mechanisms; pales of rage and death in the soul, my men struck with redoubled blows, so much so taz&hat, when I returned followed by my pack, it did not find any more on the ground, in the mud, that pieces of shapeless scrap, where, one would have had difficulty in discovering the 30 machine guns of which we would have had so much pleasure to be used!

During the interrogation that followed, Muraviev, having acquired the conviction that the French would not take sides either for or against the Bolsheviks, gave the order that we should be set free, and even that our weapons should be returned to us. The last part of this order could not be carried out, because our revolvers had already taken an unknown direction. Some of my comrades, however, were allowed to choose, among the weapons seized during the last searches, those that would suit them.

When the "capture of Loubny" was considered to be completed, i.e. when the sacking of the bourgeois houses was accomplished, the armored trains resumed their forward march: At that time, I had quite often the opportunity to be in contact with the governor of the city. He was a man with no ill intentions, who was opposed to robbery to a certain extent, and on condition that one did not forget to pay a good price. He had no political thoughts in his head, only one idea: to make a quick fortune."

As the Bolshevik army approached Kiev, the local communists there launched an insurrection, which was put down by the troops of the Ukrainian Rada then led by the former journalist Symon Petliura. But the advance of the Muraviev army could not be stopped and furious street fighting took place in the city. The head of the French air fleet, Lt Mayeur, was there and directed the offices of the French air mission under the orders of Major Bordages, assisted by some soldiers:

"*The fighting in Kiev was extremely severe, and lasted eight days. The Muraviev army was mostly composed of sailors from Kronstadt, who were fervent supporters of the revolution. I had a barricade under the window of the house where I lived, I witnessed all the fights by night and by day. One day near the doorstep I heard a complaint and I went to see, there were two Bolshevik sailors wounded. One was shot in the stomach, the other in the arm. Feeling pity, I had my men bring them back to the building where I had all the inhabitants installed in the basement. This saved our lives. Two or three days later, the barricade gave way, the Bolsheviks, drunk with rage and blood, and perhaps with vodka, rushed into the building and shot from right to left. I had my kepi pierced by a bullet. They started shouting "we are conducting searches!". They saw our entire installation in the cellar, and the two wounded Bolsheviks. The chief said to me: "Ha! You are torturing them! The sailor who had a bullet in the stomach was almost dead. The other one said: "No! They were good to us, they looked after us. We ask you to spare them! That saved the lives of me, my men, and the people in the building, a whole family of fat sugar makers who were obviously terrified. I was the king of the cellar, they didn't move a child's bed without asking me. Still, the French officers were accused of having cooperated with the White Russians. About ten of us were sentenced to death. Every night we were told that we would be shot the next morning. As for me, I had two revolvers, each loaded with seven bullets, and I told myself that I would kill ten before being shot! This idea persisted in my brain... We still had an ambassador in Moscow, who got us out of there, and we were told that we were leaving for Murmansk! They gave us a train, for all the French officers and soldiers. The Bolsheviks did not shoot us because they did not know how the situation would develop.*"

The city of Kiev fell into the hands of the Bolsheviks on February 9, 1918, but they did not hold it for long. The Germans, who had remained in their lines as spectators, noting the chaos in the Russian army, launched a vast offensive to seize the regions they had claimed in the Brest-Litovsk armistice talks, which Trotsky, the Bolshevik government's negotiator, had dragged out a little too long for their taste. They launched Operation Faustschlag on February 18, 1918, and seized the city of Kiev on February 25 without encountering any organized resistance, continuing their advance towards the Don and the Baltic countries, which they seized in the following months.

The French present in Ukraine will not give them the pleasure to wait for them. The French general Georges Tabouis, that Clémenceau had just appointed as emissary to the young Ukrainian government, ordered all the French military to leave Ukraine for Moscow. The air mission and the echelon of the Kiev park, their equipment looted, fled from the city and reached Loubny and then Moscow, not without difficulty as S/Lt Sers recalls: "*We left Ukraine on February 18, for Moscow, where all our missions were to meet. Needless to say, at each station we had to negotiate with the Bolshevik commander who always asked us the same questions and especially asked us if we did not have "polimiotes" (machine guns) of which they were very afraid. At each station, we also had to defend ourselves against the invasion of our wagons by wandering soldiers, increasingly affected by travel monomania. A detachment of our mission, which was not far from Odessa, had an incredible opportunity. A local revolutionary leader suddenly had the idea of taking the same train to Moscow, and since he had studied in France and spoke our language very well, he did his utmost to facilitate our comrades' journey. Their special train, that is to say, only for the French, at least in theory, took barely forty-eight hours to reach Moscow, whereas normally it should have taken five or six days.*"

All the French military missions, gathered in Moscow, still had to face a vileness of the Bolsheviks, according to Lieutenant Mayeur: "*They put our troops in a barracks and they were served chicken and vodka, all sorts of good things. They were brainwashed every day with suggestions to kill their officers. Our men enjoyed the preferential treatment, but the propaganda did not take hold of them! The train we were given was made up of cars with eight of us in each compartment. We stayed a whole month from Kiev to Murmansk in this train, fortunately I had plenty of food. It was cabbage soup with grains, which the Russians call kacha. We heated ourselves with stoves made of old cans, we tore off the inner wall to burn it. After a month we arrived in Murmansk, where there was cholera, and no ships, and 2 or 3 meters of snow. All the foreign nations that were evacuating their citizens were there. We lived in railroad cars, there was a huge traffic jam at the station! After eight days, an English boat arrived, used for cruises, and which could receive 300 passengers. We embarked 3000 on this boat, not only there were French troops, but there were also Czechs who had deserted from Austrian troops to serve in the Russian army. There were a thousand of them, in addition to the French. Needless to say that the comfort was basic. Finally, we landed in Aberdeen, England.*"

The troop crossed England to Southampton and finally landed in Le Havre, ending a one-year adventure for the pilots of 581 and 582 squadrons, and much more for the veterans of the first Berger mission of 1916, and even one to two years more for the volunteers such as Marcel Plat or Alphonse Poiré who were also part of the return trip.

2.4 – In the Russian civil war

2.4.1 – Lachmann's own private war

However, a distinguished airman is missing: Lieutenant Georges Lachmann, the ace of the aces of French airmen in Russia with 7 aerial victories. He was part of the trip to Murmansk, but while he was in the city of Petrozavodsk on the shore of Lake Onega (northeast of Petrograd), he was ordered to return to Moscow to take orders from the French military attaché, General Lavergne, who charged him with a mission of surveillance of the Bolsheviks' actions on the Russian military air force, which was regrouping in Moscow, having lost many aircraft (a third of its strength) in the retreat against Operation Faustschlag. Lachmann met several veterans of the Imperial Air Force who were reluctant to serve the new power. He was then hunted down by Bolshevik agents and had to leave Moscow to flee to Murmansk on June 25, 1918, but he went there with 23 Russian pilots that he managed to gather with him, including the Russian ace Alexander Kazakov, who ended the war with 18 victories.

In Murmansk, Lachmann was ordered to return to Moscow for a new liaison mission, but he finally left for Murmansk on July 10, hunted by the Bolshevik police. Arriving in Petrograd on July 11, he learned that the Bolsheviks were now in Petrozavodsk and are arresting officers, blocking the passage to Murmansk where British troops have landed in support of the white troops that are forming throughout the country. Despite this, Lachmann managed to reach Murmansk with a fellow traveler, using various means of transportation. He arrived there on August 2, 1918 and met the Allied troops.

He met the British general commanding the landed troops and proposed to him to set up a commando operation to seize the city of Petrozavodsk: "*During my arduous crossing of the Bolshevik lines, I had realized that a bold attempt could be made and had a chance of success Petrozavodsk, the most important center of the entire region, located 350 versts south of the zone of occupation by the allies, could be taken almost without casualties, provided that the railway south of this city was cut beforehand, in order to prevent the arrival of Bolshevik reinforcements. As a city of about 20,000 inhabitants, it could form a large center of resistance for us and would allow the good Russian elements to rally more easily to the Allies. I was sure, at that time, that this expedition would succeed, provided it was carried out quickly, with a very small number of troops. (...) After several days of efforts, I obtained a positive decision and a small expedition was organized. The English captain Sheppart had the command in title, I had it in fact! 24 Serbs, 14 Englishmen, 2 Frenchmen composed the staff. 2 Lewis machine guns, some explosives, rations (good English rations, ugh!) for one month. Finally, to gain speed with the Bolsheviks, several mule teams.*"

The small troop embarked on a tugboat en route to Sumskii Posad, a port on the White Sea. The ship ran aground on a sandbank and took more than a day to be cleared, while time was running out and the Bolsheviks were getting stronger... When they finally arrived at the port, the disembarkation was difficult and the little troop headed south for 100 km. But they have to stop, for lack of food, and find themselves surrounded by the Bolshevik troops in a vast landscape of swampy forest. They finally managed to free themselves and reach the coast of the White Sea on September 3, 1918, where a British trawler took them in. Lachmann went to the city of Arkhangelsk, a port on the White Sea at the mouth of the Dvina River where the British troops had landed. He testified about the rest of his journey:

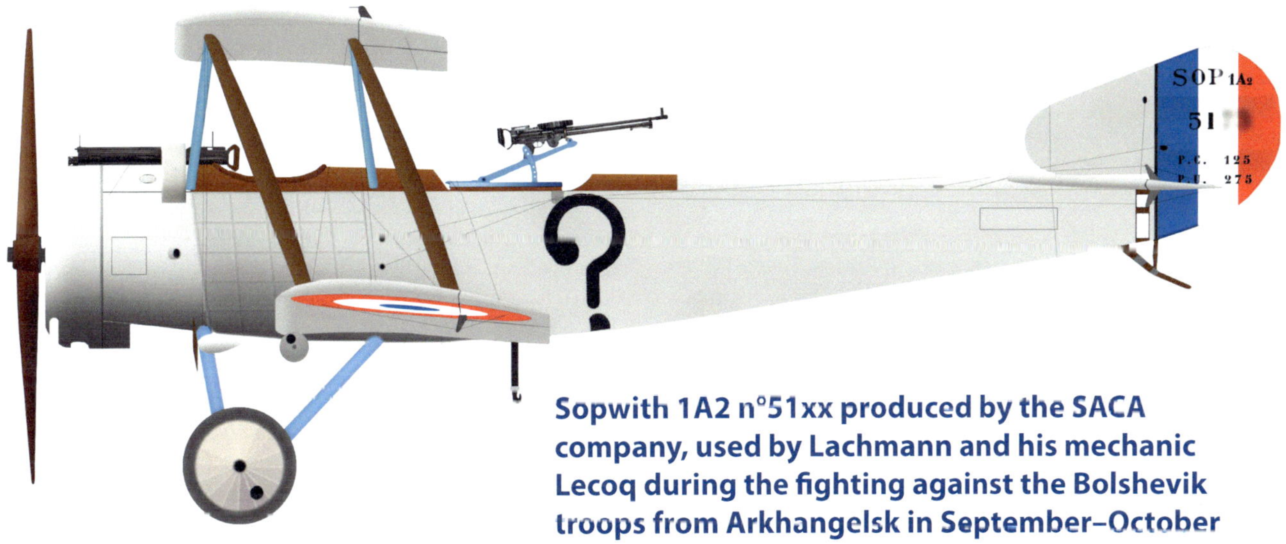

Sopwith 1A2 n°51xx produced by the SACA company, used by Lachmann and his mechanic Lecoq during the fighting against the Bolshevik troops from Arkhangelsk in September–October 1918, within the Slavo-British Aviation Corps (SBAC)

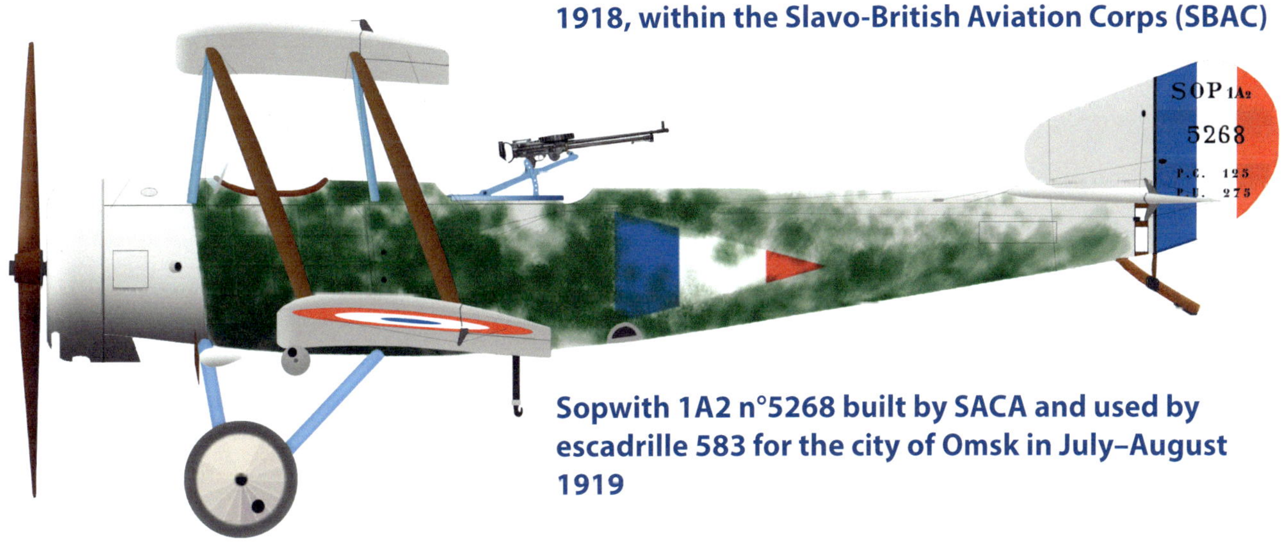

Sopwith 1A2 n°5268 built by SACA and used by escadrille 583 for the city of Omsk in July–August 1919

"In Arkhangelsk, I found Colonel Maund, who was in command of the British air force in Galicia on the former Russian front, as commander of the Allied air force. I was his deputy. An English squadron was operating with the Allied forces along the Arkhangelsk-Vologda railroad line. On the Dvina, another part of the front, we organized two squadrons with the Russian pilots that I had sent from Moscow in time. Kazakov, the Russian ace, was in command: the first one left at the beginning of September, the second one on the 10th. The aircraft were the 18 Sopwiths, the only Allied aircraft found in Arkhangelsk, all the others having been taken in time by the Bolsheviks. Alas! Poor squadrons! Wonderful pilots of courage and spirit. Flying for hours along the railroad or the river, always over the woods or swamps, without a field (in one month, 3 planes of the English squadron landed on the forests by breakdowns). For 20 pilots there were 6 mechanics, all the rest of the personnel recruited in Arkhangelsk; in the second squadron, there were 18 Chinese, former stevedores of the port! These were the famous Slavo-British troops raised in Arkhangelsk.

I left at the same time as the second squadron with a Sopwith equipped for bombing and my faithful mechanic Lecoq, called Coco. All the old airmen knew him: he spoke little, was surly, and only had fun when I took him to bomb the Bolshevik fleet on the Dvina River. He sank two large ships there, quite nicely. It will not seem so strange when I say that the Dvina is more than 2 kilometers wide in some places, and that the English monitors could sail up this river for 350 kilometers. At that time the Bolsheviks had 21 ships on the river, some of them carrying several 150 mm cannons. I was very surprised when I was surrounded by special batteries!

Soon the winter came, the first snows. No hangars, no tents! No new equipment, no spare parts! Another month and 25° below zero would prevail for days of two hours of daylight. I decided to return to France, to finish my dozen Krauts. To reach Arkhangelsk, two and a half days by boat on the Drina (twelve days for the climb!). I am happy to find myself one of the first who flew under these latitudes for long hours, over desolate lands, deserts, woods or swamps. I often think of one of these, of more than 100 square kilometers, that I flew over every day, sometimes several times, to go from Chenkoursk, where I was then, to the Dvina where the Bolsheviks were. An immense black hole, of more than 45 kilometers in the center, seemed to me the immense crater of some volcano. Crater of mud, from where, in case of breakdown, my 8 days of food that I always had in the plane, would not have taken me out.

In Arkhangelsk, a pleasant surprise, the British Military Cross was awarded to me: "From August 8 to September 6, 1918, Lieutenant Lachmann was second in command of a reconnaissance detachment that penetrated the enemy lines south of Sumskii Posad. It was largely due to his courage and drive that the enemy, who was far superior in numbers, was repulsed on several occasions with considerable losses. He showed the greatest disregard for danger, on several occasions leading his men in the most courageous way, in conditions of inequality, thus giving a splendid example during the whole time of a very risky enterprise.

This proves that the aviators who have always been only aviators, are as good infantrymen as the others, whatever some people say! I embarked on November 7, 1918, eager to meet the Kraut birds again. The news of the armistice surprised me (oh so much!) in a small Norwegian fjord, halfway between the North Cape and Bergen.

Thus ended the French aviation in Russia, its last representative having failed to get out!"

2.4.2 – Last flights of the French on the Trans-Siberian Railway: the short-lived 583 Squadron

While Lachmann was fighting the Bolsheviks in the region of Arkhangelsk with the Allied troops led by the British, another front had been set up against the Bolsheviks along the Trans-Siberian Railway, the only means of communication through Siberia, running eastward to Vladivostok.

The signing of the Treaty of Brest-Litovsk resulted in the repatriation of all German, Austro-Hungarian, and Turkish prisoners of war to their respective countries: railway convoys travelled from east to west loaded with these prisoners. At the same time, soldiers of the Czech Legion, a military unit of nearly 65,000 men made up of former Austro-Hungarian prisoners of war of Czech and Slovak nationality, wanted to continue the struggle on the side of the Allies and had to be evacuated to Vladivostok. The Trans-Siberian Railway was very crowded with convoys running in both directions and the Czech legionnaires became impatient... On May 14, 1918, an incident broke out in Chelyabinsk: an altercation between Hungarian prisoners and Czech legionnaires degenerated and a Hungarian was killed. The local Bolshevik authorities arrested the Czech legionaries involved in the incident and ordered their execution. The entire Czech legion in the city rose up to free its comrades and swept away the Bolshevik authorities. Other Czech legion detachments posted along the Trans-Siberian Railway did the same; they were joined by local garrisons led by white officers. Within a month, the entire Trans-Siberian Railway, from the Ural Mountains to Lake Baikal, came under the control of the whites, who set up several local governments, most of them headed by reformist socialists, especially in Samara, which was captured by the Czech Legion on June 8, 1918.

The White forces, spearheaded by the Czechs, managed to seize the city of Kazan on August 7, 1918, where the gold reserves of the Russian Empire were captured by the legionnaires. Lenin's government could not remain without reacting to the capture of Kazan when the next stop of the Trans-Siberian Railway was the city of Moscow. He ordered the 5th Red Army, which had to leave the city, to counter-attack and allocated substantial reinforcements, including a military aviation that they formed with the remains of the squadrons of the Tsarist aviation that had survived the Faustschlag operation, reinforced with some aircraft from the Duks factory in Moscow, which still had stocks of spare parts to produce new aircraft.

This air force was directly commanded by the head of the Bolshevik air force himself, Comrade Konstantin Akashev, who knew that he was playing for time... He had his work cut out for him, as many pilots of the former Tsarist air force were not ardent supporters of the new regime and took advantage of the arrival of the White troops to leave and join them with their aircraft. This is the case of two Franco-Russian pilots, starting with Henri Bourrier, former chief of the 23rd KAO who deserted in August 1918. He was followed by Jean de Voyod who deserted on his Nieuport 17 with one of his comrades on August 19, 1918 – they landed on the Kazan airfield that they were supposed to bomb. All of them then served in the White Army of Admiral Kolchak, who in November 1918 became the leader of the entire White Army of Russia.

From then on, he would work to reorganize the White Army, with the support of the British who would deliver him weapons and ammunition. However, Kolchak showed serious shortcomings in the role of the air force, and did not take any immediate initiative to strengthen his meager air force, which was still flying on the aircraft seized the previous year and which had undergone many repairs. It was more or less blindly, without systematic reconnaissance, that his troops went on the offensive again against the Red positions at the beginning of 1919, approaching the cities of Kazan (which the Red Army had taken back on September 10, 1918) and Samara, but without succeeding in taking them. Kolchak's military machine began to seize up: the soldiers of the Czech Legion, learning that their country had obtained its independence, tried to return and gradually withdrew to Vladivostok, depending on the railroad convoys available.

However, in July 1919, Kolchak saw an appreciable air reinforcement arrive in the city of Omsk by the Trans-Siberian Railway, the French squadron SOP 583, which had to travel around the world to reach its destination. It was led by a connoisseur of Russia, Captain Henri Moutach, a veteran pilot of the Berger mission of 1916, who obviously has a serious resentment against the Bolsheviks! With him came eight other pilots (Capt. Roland Thévenot, Lt. Maurice Amet, Lt. Henri Tresch, S/Lt. Max Bruyère, Adj. Albert Perrot, Sgt. Francis Roubey, Sgt. Henri Grevedon, Sgt. Rodolphe Calvignac) and as many Sopwith 1A2s intended for reconnaissance and artillery adjustment.

But the problem for Kolchak was that this squadron had

Above: Three gentlemen having a drink in a club in the Shanghai International Concession in 1941... Surrounding the British Squadron Leader Mc Bain are two "French" veterans of the Russian Imperial Air Force, former Colonel Léon Ducimetière (left) and Captain Jean de Voyod (right). (DR)

Below: Ukrainian soldiers mounting an honor guard in 2019 in front of Marc Bonnier's grave at the Odessa cemetery (DR).

Riga
Pskov
Memel
Dunaburg
Kaunas
Danzig
Konigsberg
Vilnius
Minsk
GERMANY
Grodno
Tannenberg
RUSSIAN EMPIRE
Warsaw
Brest-Litovsk
Pinsk
Lodz
Breslau
Radom
Lublin
Kovel
Loutsk
Rowno
Kiev
Dubno
Tarnow
Krakow
Przemysl
Lvov
Gorlice
Tarnopol
Vienna
Czernowitz
FRONTLINE LATE 1916
FRONTLINE LATE 1917
Budapest
EMPIRE AUSTRO HONGROIS
Jassy
0
100
200
300 km
Galatz
ROMANIA

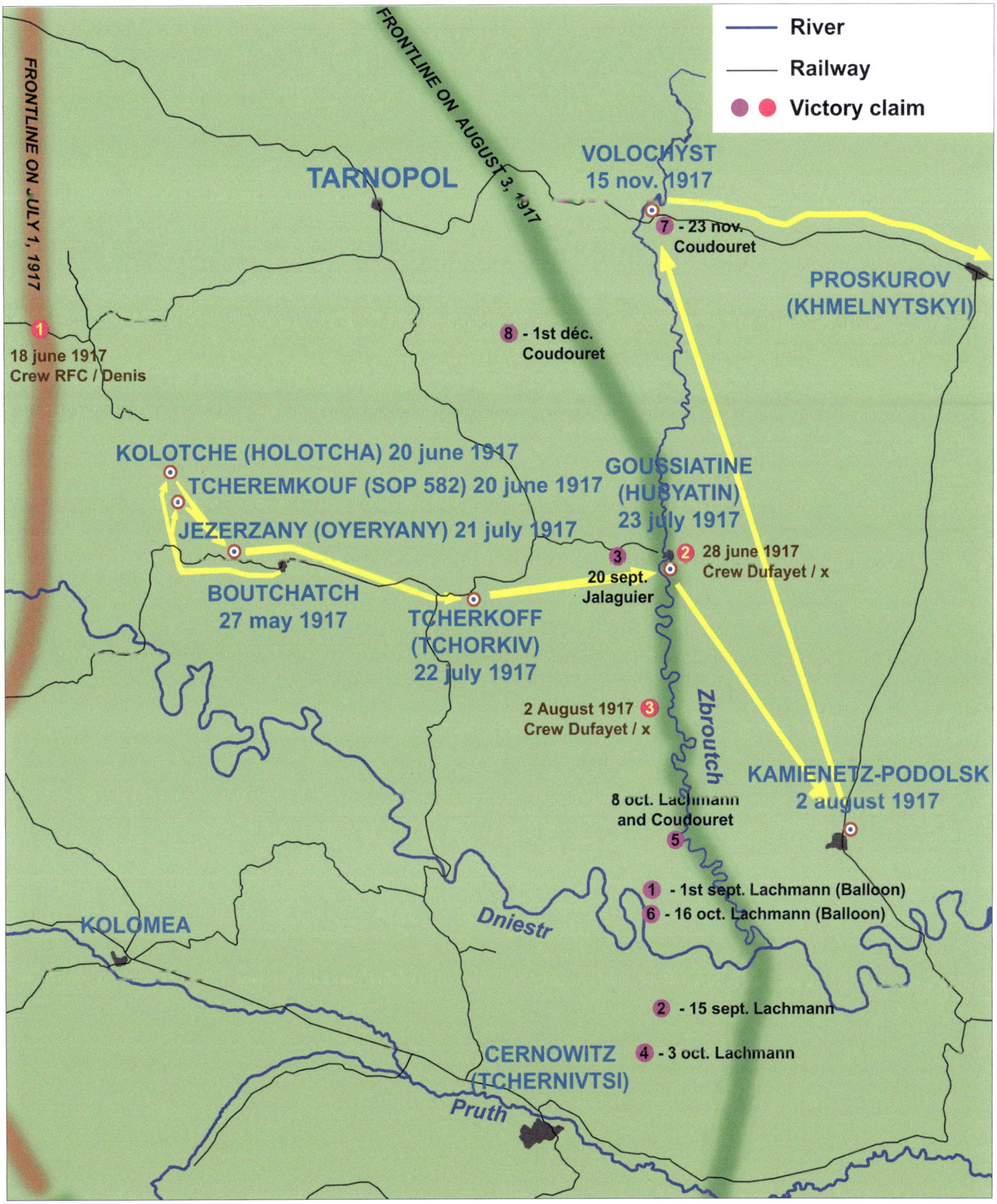
River
Railway
Victory claim
FRONTLINE ON JULY 1, 1917
FRONTLINE ON AUGUST 3, 1917
TARNOPOL
VOLOCHYST
15 nov. 1917
7 - 23 nov.
Coudouret
PROSKUROV
(KHMELNYTSKYI)
1
18 june 1917
Crew RFC / Denis
8 - 1st déc.
Coudouret
KOLOTCHE (HOLOTCHA) 20 june 1917
TCHEREMKOUF (SOP 582) 20 june 1917
JEZERZANY (OYERYANY) 21 july 1917
GOUSSIATINE
(HUSYATIN)
23 july 1917
3
20 sept.
Jalaguier
2 28 june 1917
Crew Dufayet / x
BOUTCHATCH
27 may 1917
TCHERKOFF
(TCHORKIV)
22 july 1917
2 August 1917 3
Crew Dufayet / x
Zbroutch
KAMIENETZ-PODOLSK
2 august 1917
8 oct. Lachmann
and Coudouret
5
1 - 1st sept. Lachmann (Balloon)
6 - 16 oct. Lachmann (Balloon)
KOLOMEA
Dniestr
2 - 15 sept. Lachmann
CERNOWITZ
(TCHERNIVTSI)
4 - 3 oct. Lachmann
Pruth

Victory Claims on Russian Front by French Pilots or Observers

#	Date	Rank	Pilot or Gunner		Unit	Plane Piloted	Plane Claimed
-	3/10/15	Poruchik	Sebalev, S.K.	-	2 AAO	Farman ?	Airplane (FTL)
		Adj	Poiré, Alphonse	-	2 AAO		
1	7/1/16	S/Lt	Pulpe, Edward	4	8 AOI	Nieuport 11	Airplane
-	?/11/1916	Feldwebel	Plat, Marcel (gunner)	-	EVK	Ilya Murometz	Airplane (FTL)
2	3/22/17	Lt	Paillette, Marcel	1	6 AOI	Nieuport 11	Airplane
3	3/22/17	Lt	Paillette, Marcel	2	6 AOI	Nieuport 11	Airplane
4	6/18/17		RFC pilot		RFC	BE 2	Airplane
		Lt	Denis, Maurice (observer)	1	SOP 582		
5	6/28/17	S/Lt	Dufayet, Pierre	1	SOP 582	Sopwith 1A2	Airplane
			Unknown observer				
-	7/22/17	Lt	Smith (RFC pilot)		RFC	Sopwith 1A2	Aiplane
		Lt	Denis, Maurice (observer)	2	SOP 582		
6	8/2/17	S/Lt	Dufayet, Pierre	2	SOP 582	Sopwith 1A2	Airplane
			Unknown observer				
7	9/3/17	S/Lt	Lachmann, Georges	3	N 581	SPAD VII	Drachen
8	9/15/17	S/Lt	Lachmann, Georges	4	N 581	SPAD VII	Two-seater
9	9/20/17	Adj	Jalaguier, Paul	1	N 581	Nieuport	Airplane
10	10/3/17	S/Lt	Lachmann, Georges	5	N 581	SPAD VII	Brandenburg C I 169.14
11	10/8/17	S/Lt	Couduret, Louis	3	N 581	SPAD VII	Albatros C
		S/Lt	Lachmann, Georges	6	N 581	Nieuport	
12	10/16/17	S/Lt	Lachmann, Georges	7	N 581	SPAD VII	Drachen

1914–1919

Place	Source	Confirmation	Notes
Vovicz		None	
		None	
Lutsk / Kovel	Squadron report	Army report	
?	Report of Murometz's pilot, Lt Lavrov	None	
Volodymyr-Volynskyï (W. Lutsk)	Russian communiqué	Citation OA 18-03-1919 - JORF 30-05-1919 p.5564 "Shot down 2 planes"	
Volodymyr-Volynskyï (W. Lutsk)	Russian communiqué	Citation OA 18-03-1919 - JORF 30-05-1919 p.5564 "Shot down 2 planes"	
Michischuv railway station			Attacked by two airplanes, shot down one in enemy lines
		Citation OA 20-11-1917 - JORF 22-01-1918 p.812	
Gusyatin	Russian communiqué		
Zalec'ye, near Zbrousch river		None.	Airplane from Flik 13 damaged in sector, observer wounded
Skala / Losich (NO Losich)		Dufayet received St Georges order, 4rd class	During a combat with 2 enemy planes
20 miles from Melnitza, near Chotin	Russian communiqué front 12th Army Corps, 18-10-1917.	Confirmed by next citations	Ballonkompagnie 22 (Kuk) - LdR Egon Wertheimer et Fuerwerker Frantz Antos escaped in parachute
Chilovsti (Chilovtse)	Communiqué 7rd Russian Army 18-09-1917	Citation OA 30-10-1918 (1st) - JORF 30-12-1918 p.11261 +	
Chabaruvka	Communiqué 7rd Russian Army		
Mal'Shevka (Malyshevka-Malistoni)	Communiqué 7rd Russian Army 04-10-1917	Citation OA 30-10-1918 (1st) - JORF 30-12-1918 p.11261 "5th victory" +	FlAbt 242w, plane damaged
Zalec'ye, near Zbrousch river	Russian report n°254	Confirmed by next citations	
25 km W. Kameniez - Poldolsk	Communiqué 3rd Russian Army Corps 08-10-1917	Citation OA 30-10-1918 (2nd) - JORF 30-12-1918 p.11261 "6th and 7th victoires"	
Melnitza (18h20)	Communiqué Russian 12th Army Corps 18-10-1917 "Second enemy Drachen"	Citation OA 30-10-1918 (2nd) - JORF 30-12-1918 p.11261 "6th and 7th victories"	Ballonkompagnie 22 (Kuk) - LdR Egon Wertheimer et Fuerwerker Frantz Antos escaped in parachute (a second time after September 3...)

Victory Claims on Russian Front by French Pilots or Observers

#	Date	Rank	Pilot or Gunner		Unit	Plane Piloted	Plane Claimed
13	11/23/17	S/Lt	Couduret, Louis	4	N 581	SPAD VII	Albatros C X
14	12/1/17	S/Lt	Couduret, Louis	5	N 581	SPAD VII	Albatros D V

French Losses on the Russian Front 1914–1919

Date		Pilot (P) or Observer (O), Fate	Unit	Plane Piloted
6/10/16	Praporshchik S/Lt	Shvarts, Bertold Ya. (Pilot, POW) - RUSSIAN Cotier, Jean (Observer, POW)	32 KAO	Voisin
6/12/16	Lt Sol	Méritel, François (Pilot, KIAcc) Bouvrier, Maurice (Observer, KIAcc)	27 KAO 27 KAO	Morane
8/2/16	S/Lt	Pulpe, Eduard Martynovitch (KIA) - RUSSIAN IN FRENCH ARMY	8 FAO	Nieuport 11
8/9/16	S/Lt	Bonnier, Marc (KIA)	Anatra factory	Anatra D n°190
9/5/16	Praporshchik Sgt	Pankratov, Onisim Petrovich (Pilot, KIA) - RUSSIAN Laurent, Emile (Observer, KIA)	5 FAO 5 FAO	Nieuport 10 n°711
3/2/17	Praporshchik S/Lt	Grechkin (Pilot, WIA) - RUSSIAN Berné, Raymond (Observer, KIA)	2 AAO 2 AAO	Voisin-Lebedev n°422
6/22/17	Adj-Ch	Lignac (-de), Louis (WIA)	?	Farman
7/23/17	S/Lt	Robinet, Jean (Pilot, KIAcc) Omelin, Ivan (Observer, KIAcc) (RUSSIAN)	Anatra factory	Anatra DS
11/26/17	MdL	Chandon, Lucien (KIAcc)	Anatra factory	Anatra

1914–1919

Place	Source	Confirmation	Notes
Near Volochyst		Citation OA 30-10-1818 - JORF 30-12-1918 p.11261 "4th victory"	Fl Abt (A) 232, crew captured
Sorotska, SW Skalat	Communiqué Russian 7th Army	Citation OA(LH) 30-05-1918 - JORF 09-06-1918 p.4963 "Shot down 5 enemy aircraft, including 3 in Russia"	

Cause	Place	Notes
Air combat + Flak	Monasterzhiska, near Podgaitsy	Pilot escaped on 09/01/1917
Accident	Akimofisha (near Minsk)	
Air combat	Roshzhishche (Kovel / Lutsk)	Air combat against 3 Fokker
Engine malfuction		
Air combat	Lake Drysvatsy (5th Army sector, North front)	
Air combat	SW Baranovichi (near Lyakhovichi)	
Flak	?	DOW 07/07/1917
Accident	Odessa	
Accident	Tchoulaicjka, near Odessa	

been created by the French government to serve as the air component of the Czech Legion. This one folded up, the 583 squadron had to do the same as soon as it arrived... Three French pilots (Lt Amet, S/Lt Bruyère, and Sgt Calvignac) were nevertheless going to carry out reconnaissance missions for about a month on their own equipment in the White Russian units. They surprised Kolchak's airmen by their efficiency and the fact that their planes did not break down... This made the Kolchak government aware of the aerial fact. He bought the Sopwith from the French of the 583 squadron which left the country at the end of August 1919 for the last of them. The pilots left Russia and left Kolchak to his fate (executed on February 7, 1920), putting an end to five years of French military air presence in Russia.

23 years later, on November 29, 1942, fourteen French fighter pilots and 58 mechanics set foot on Russian soil to form the Normandie squadron. But this is another story...

Sources:

- SHD, box A 189 - French military mission in Russia
- Lt Emile Stribick, Berger mission pilot: "The air war among the Bolsheviks", in "La guerre aérienne illustrée" of 1918
- Lt Georges Lachmann, squadron leader 581 : "My campaigns in Russia", in "La vie aérienne illustrée" of 1919.
- Lt Louis Coudouret, pilot of the 581 squadron: "French aviation in Russia", in "La guerre aérienne illustrée" of 1918.
- Lt Georges Mayeur (officer of the park) : oral testimony SHD n°284
- Lt Louis Sers (deputy park officer) : "Un parc d'aviation française en Russie bolchéviste mars 1917 - mars 1918", published in the Revue des deux mondes tome 46, in 1918 - available on Gallica and Wikipedia
- Lt Jules Legras, officer of the 2nd office in Russia : memoirs, Dijon municipal library.
- Log book of escadrille 583, SHD.

Chapter 3: French Military Mission in Romania 1916–1918

Above: A Morane Parasol of the Romanian Army Air Corps (Corpul de Aviație) takes part in a military parade in May 1916, before the country entered the war, which had about forty aircraft, half of which were in flying condition (SHD).

3.1 – A highly courted kingdom

The Kingdom of Romania was formed in the second half of the 19th century at the expense of the Ottoman Empire, through the union of two of its provinces, Wallachia and Moldavia. However, many of the Romanian-speaking population remained under the sovereignty of the two other great empires of the region, Austria-Hungary, which possessed Transylvania, and Russia, which possessed Bessarabia, that is to say the part of Moldavia situated between the rivers Pruth and Dniestr. At the time of the first Balkan war in 1912, when the states of the region (Serbia, Greece, Montenegro, Bulgaria) attacked the remnants of the European possessions of the Ottoman Empire, Romania remained cautiously neutral, but it was to become involved in the second war of 1913, when the above-mentioned countries joined forces against Bulgaria. In this second conflict, Romania won the province of Dobroudja, south of the mouth of the Danube, populated only by a minority of Romanians.

When the First World War began, the Romanian government, following the opinion of its King Carol, chose the path of neutrality. But the entry of Bulgaria into the war in 1915 and the invasion of Serbia, followed by the presence of French troops in the north of Greece where the Eastern Front was formed, changed the situation in the region. Both sides of the war actively courted the Romanians: the central empires promised Bessarabia, while the Entente diplomacy led by the Minister of Foreign Affairs and then President of the Council Aristide Briand promised him Austro-Hungarian Transylvania and Bucovina. A skilled negotiator, Briand was convincing and won the support of the Romanian government as well as that of the new King Ferdinand I, whose wife, Queen Marie of Romania, of British origin, had actively campaigned to join the Entente. A political agreement was signed on 17 August 1916, supplemented by a military agreement on the 25th.

Even before the agreements were signed, France sent a discreet military aviation mission of three pilots to Romania,

Above: Portrait of Căpitan aviator Vasile Craiu, one of the few Romanian military pilots before the country entered the war. He was described by French pilot René Chambe as an admirer of Germany... (DR)

composed of Lieutenant Jules de Baillardel de Lareinty-Tholozan, Lieutenant René Chambe, and MdL Gilbert Adam, who set out for Romania to train local aviators in the piloting of the few Nieuport 11 (so-called "13-meter") fighters and Farman F.40s that were also sent by the French government via Russia. The three French pilots, accompanied by mechanics Gabriel Billet and Louis Derozier, embarked for England dressed as civilians, from where they reached Norway (Chambe arrived there on August 12) and then Sweden, two Scandinavian countries that were neutral at the time, and entered Finland, which belonged to the Russian Empire. Traveling by train and stopping in Moscow and Kiev, they were able to enter Romania and settle in Bucharest on August 25 for Chambe and Adam, de Lareinty-Tholozan having preceded them by a week. René Chambe, who became a prolific writer on his long military life, left a detailed account of his trip in 1981 ("Route sans horizon - Les eaux sanglantes du beau Danube bleu") and marveled at the Russian army, which he judged to be very disciplined, as far as he could observe from the few detachments he came across during his trip. The French ambassador, Mr. Maurice Paléologue, whom he met in Moscow, was anxious to temper his enthusiasm and informed him of the critical state of the Russian Empire, which he knew to be severely tested by the war. The future help of Russia to Romania would undoubtedly be problematic, because of the territorial dispute concerning the region of Bessarabia.

Upon their arrival in Romania, the French discovered in Bucharest the local air force, which they considered dramatically insufficient. The Romanian air force had had its baptism of fire during the Second Balkan War of 1913, when its few Bleriot XIs carried out reconnaissance on Bulgarian enemy positions. In January 1915, a purchase mission was entrusted to Prince Georges Valentin Bibesco, president of the Romanian National Air League, who acquired some forty aircraft in Italy, delivered via Salonika on July 9, 1915, before Italy entered the war and the port of Salonika was occupied by Allied troops. Almost all of the purchased aircraft were of French design; one year later, in August 1916, Romanian aviation had 24 aircraft in flying condition out of the 44 listed in the official inventory: 8 Maurice Farman MF XI, 5 Blériot XI, 4 Morane Saulnier Parasols LA, 4 Voisin L, 1 Henri Farman HF 20, 1 Caudron G.3 and one Aviatik. There were 97 licensed military aviators to fly them.

According to Chambe and de Lareinty-Tholozan, this was not enough to carry out the necessary reconnaissance for the

Voisin LA of the Corpul de Aviatie in 1916.

Above: Hansa Brandenburg C.I two-seater of the Austro-Hungarian aviation, of which it is the real workhorse because it would remain on the line during the whole war, being improved by the installation of more powerful engines. (Coll. B. Toschinger)

armies; moreover, there were no fighter aircraft. They wrote a report that the French military attaché in Bucharest, Colonel Desprès, did not see fit to transmit, especially since a first batch of 18 Nieuport 11 fighters arrived by train. Chambe brought up the first of them, bearing the serial number 1405, and was able to give a flight demonstration to the Romanian aviators... But the aircraft, like the others, was not armed with its machine gun.

3.2 – A quick military defeat

Romania officially entered the conflict on August 27, 1916, declaring war on Austria-Hungary and attacking the Transylvania region, leaving a curtain of troops facing Bulgaria along the Danube and the Drobudja. The Romanian government hoped to count on Bulgarian neutrality, which it did not obtain because in the days that followed this country as well as Germany and the Ottoman Empire declared war in their turn.

To the great despair of René Chambe, it was with a Romanian air force that was not very well prepared and did not wait for the French reinforcements that the hostilities started. The distribution of these aircraft among the four Romanian armies clearly illustrated the priorities given by the high command: the 1st, 2nd, and 4th armies, which led the offensive in Transylvania, were assigned the bulk of the air resources in the form of three squadrons based respectively at Talmaciu, Baicoi, and Pietra Neamt. Only 9 aircraft based around Bucharest remained in the south to defend the capital and monitor the Danube front.

The Romanian troops quickly crossed the Transylvanian passes and penetrated into enemy territory, supported by their small air force which carried out reconnaissance and a few bombing missions on the enemy communication nodes. But if the Romanian troops progress, it is because there were hardly any enemy troops to counter them for the moment, the Austro-Hungarian army remaining in retreat.

For the Romanian attack was anything but a surprise for the Central Empires, who had already organized themselves under the leadership of Germany. The negotiations with the Entente allies for Romania's entry into the war were perfectly followed by German espionage. On the Eastern Front, General Sarrail was ordered to conduct a generalized offensive on August 20, 1916, seven days before the planned attack of the Romanian troops, but the German-Bulgarian forces did not give him the time to do so and took the initiative of the assault on August 17, 1916, the Eastern Front line freezing rather quickly on positions that would remain unchanged until the end of the war. On the Romanian front in Transylvania, the 9th German army, composed of valuable troops and led by General Erich von Falkenhayn, organized a vigorous counter-attack on September 25, 1916, and in less than three weeks brought the Romanian army back to the Carpathian passes, before pushing them back beyond and invading Wallachia. The troops of von Falkenhayn were supported by the Austro-Hungarian air force, composed of five Fliegerkompanie (Flik) withdrawn from the Italian front in September, and bearing the numbers 13, 31, 32, reinforced in September by the numbers 33 and 36. Composed of two-seater Hansa-Brandenburg C.I and Lloyd C.III accompanied by some fighters, these units represent a fleet of about 80 aircraft, clearly superior to the Romanian aviation which would have to undergo the bombardment of its ground of Pietra Neamt by the enemy aircraft.

But the German-Austrian counter-attack on the Carpathians was not even the main threat that Romania

Above & Below: The Zeppelin LZ 86 (LZ 56 according to the manufacturer's serial number) leaves its hangar of Szentandrás (currently Sânandrei, near Timisoara in modern Romania) on September 3, 1916, to carry out a raid against the city of Ploesti and its oil wells. It flew a total of 7 missions in the Balkans, before being destroyed in a handling accident while docked in its hangar on September 5, 1916. A number of missions in the average of the other Zeppelins... (Coll. Jörg Biber)

Above & Following Page: Zeppelin LZ 81 (LZ 51 n° builder), seen here coming out of its hangar of Szentandrás, was not so lucky as it was hit by Romanian flak and was forced to land in the wild on September 27, 1916 at Veliko Tarnovo (Bulgaria) while returning from a raid on Bucharest. Too damaged, it had to be dismantled (Coll Jörg Biber)

LZ81

Above: View of a Rumpler G.II in flight. These bombers, much more efficient than the Zeppelins, multiplied the destructive missions during the autumn 1916 campaign which saw the conquest of two thirds of the Romanian terrıtory. (Jack Herrls)

had to face: a coalition of troops from the Central Empires, composed of the 3rd Bulgarian army, the 6th Ottoman army corps, and the German Danube army attacked from the south from Bulgaria. The Romanian troops could count on the natural barrier offered by the very wide Danube River, which served as a border with Bulgaria, but this was not the case in the Dobroudja region, where the enemy troops were concentrated, and on September 4, 1916, they won an important victory at Bazargic (today Dobritch). The Romanian troops suffered heavy losses and, despite some reinforcements received from the Russian army, had to retreat towards the Danube.

This attack from the south was supported by considerable

Right: Lt René Chambe, photographed here on the Pipera field near Bucharest, was alone in his attempt to oppose the incursions of the enemy bombers attacking the capital. He is seen here on his Nieuport 11 n°1405 which he decorated with the blue and white pennant of escadrille N 12 squadron, his former unit. The photo was published in the Romanian press which reported that he managed to disperse a formation of enemy bombers (Coll. Emmanuel de Vachon)

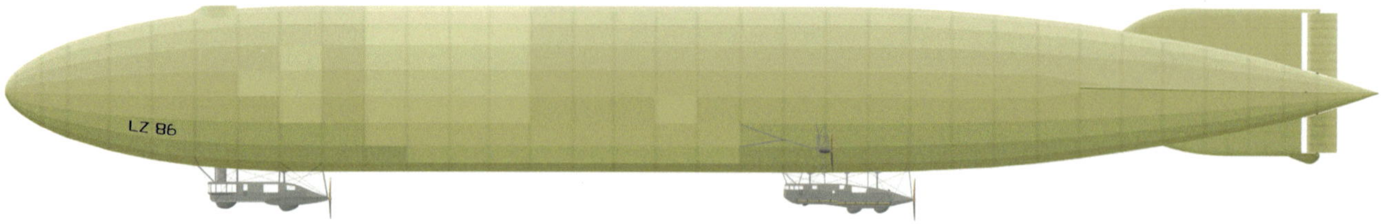

Zeppelin LZ 86 (LZ 53 manufacturer number), Szantandras, early September 1916.

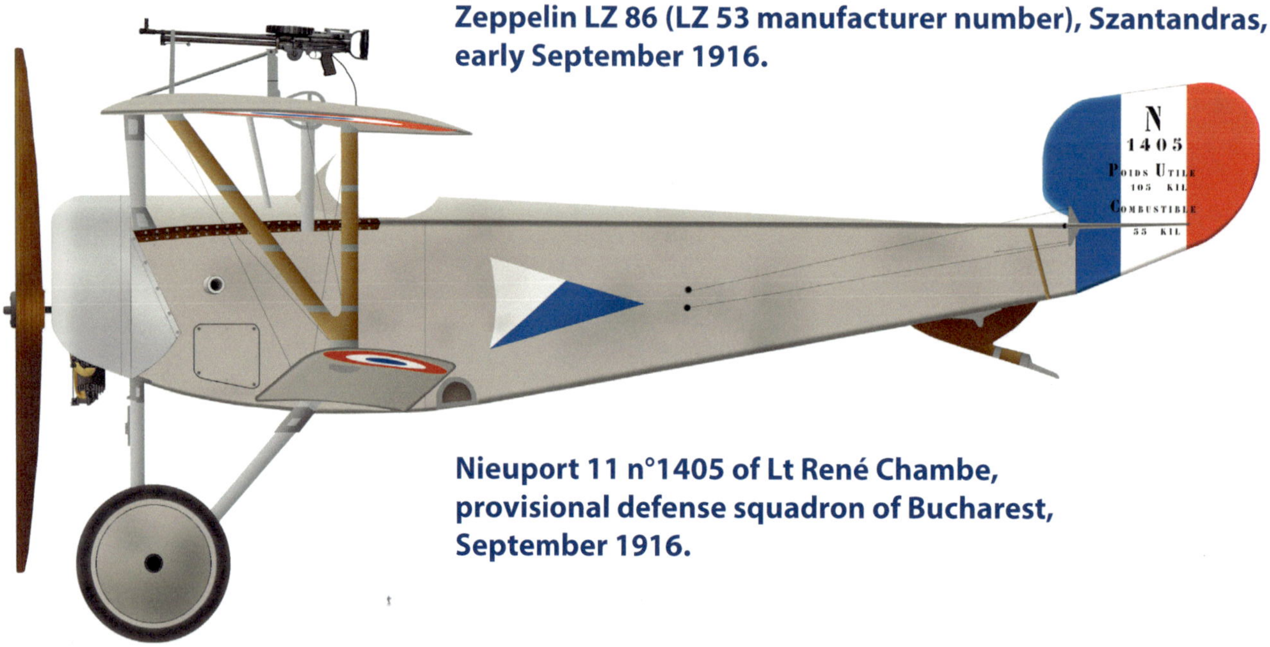

Nieuport 11 n°1405 of Lt René Chambe, provisional defense squadron of Bucharest, September 1916.

German air forces. As soon as Romania entered the war, its capital was attacked by a Zeppelin during the night of 28-29 August 1916. It was the LZ 101 (manufacturer number LZ 71) based in Yambol in Bulgaria, some 200 kilometers from Bucharest. It was able to carry out its mission without any problems: the city is not defended by the flak! The Romanians would have to improvise an air defense but would undergo other attacks of terror on their capital, again visited during the night of September 4 to 5 by the LZ 101 at the same time as the city of Ploesti which the LZ 86 (n° constructor LZ 56) attacked from its base of Szentandras in Hungary (currently Sanadrei close to Timisoara in Romania) and which, damaged during the raid, was accidentally destroyed during the docking maneuvers on the way back to its base, at the cost of the lives of 9 crew members. The Zeppelins, whose destruction was more psychological than military, remain fragile machines: a new raid was carried out on Bucharest on September 14 by the LZ 101, which succeeded the LZ 81 (manufacturer's number LZ 51) left Szentandras during the night of 24 to 25 September 1916. The two airships attacked the capital together during the following night but LZ 81 was hit and lost gas on the way back, having to land in the wild in Bulgaria at Turnovo where, damaged, it would be dismantled. Despite the loss of two airships based in Szentandras, the Germans installed on this same base the LZ 97 (manufacturer number LZ 67) which would carry out with the LZ 101 two new attacks on Bucharest during the nights of October 23 to 24 and November 1 to 2. Other objectives will be targeted by the Zeppelins, such as the port city of Feteşti, on the Danube, attacked during the night of October 24 to 25 by LZ 97 and LZ 101 while the latter would attack during other raids the cities of Ciulniţa, Galaţi, and Iaşi, until December 1916.

The Zeppelins were only one component of the German air force, which in Bulgaria had an entire squadron of bombing planes, the Kampfgeschwader I, installed in early September at the edge of the Drobouja over the city of Razgrad. The squadron was composed of three squadrons (Kampfstaffel or Kasta bearing the numbers 2, 3, and 5), plus an independent Kasta bearing the number 20. These units were mainly equipped with twin-engine bombers (mainly AEG G.III but also some Gotha G.II and Rumpler G.II), as well as some single-engine two-seater Albatros C. All these bombers were used in a grouped manner on strategic and tactical objectives and were more efficient than the Zeppelins, if only because of the frequency of their attacks: they carried out more than thirty bombing missions during the campaign, the first one during the night of 20–21 September 1916 against the city of Cernavoda, located on the Danube, where there was a strategic bridge and a railway station. Both were attacked six times but the bridge remained intact because the bombs used were too small. Bucharest and its suburbs were also among the targets of the bombers who dropped their bombs on 25 and 26 September in conjunction with those of the Zeppelins, and returned there ten more times until 12 December. They also attacked other objectives such as the coastal cities of Mangalia and Constanza where the city of Feteşti which also has a bridge over the Danube as an extension of the one at Cernavoda.

Bombers and Zeppelins could act in relative impunity, due to the lack of air opposition: the small Romanian air force was quickly overwhelmed and did not have interceptors to try to oppose these raids – an improvised patrol was mounted in Bucharest on Nieuport 11 fighters and tried to oppose the German bombers. Lieutenant Chambe was one of them and testified that he had at least succeeded in disorganizing the enemy formation and forcing it to disperse its bombs in the wilderness – with relative success, as the raid of 25 September 1916 against Bucharest killed nearly 485 civilians

3.3 – French and British aid sent in hurry

Faced with the impending disaster, the French and British forces tried to provide emergency military aid to the Romanians by air. On September 14, 1916, an air link was successfully established by Lt Noël, a specialist in long-distance flights, between Salonika and Romania in the company of Lt Leseur, who had taken the place of the observer in his Farman. The next day, September 15, two other Farmans left Salonika (Lt. pilot Quillery + machine gunner Crouzier, MdL pilot Rouen + Sgt. machine gunner Renier) and landed in Romania with their aircraft, but not without having dropped bombs on Sofia during their trip. As soon as they left for Salonika, these raids, in addition to raising the morale of the Romanians, demonstrated that it was possible to establish air links between the Army of the East and Romania, and it was by this means that the transfer of the men and aircraft of the aviation mission for Romania, which was being set up in France, was initially envisaged.

Indeed, on September 23, 1916, the Minister of War, General Roques, appointed Captain de Malherbe, a very experienced aviator (military pilot's license no. 2 obtained in February 1911), as head of this mission. He had to fight against General Joffre, head of the GQG, who did not recognize him as a commander and proposed his protégé, the artillery captain Emile Pierra, but Roques maintained his decision. De Malherbe gathered around him a total of 283 men of all specialties, including 20 Nieuport fighter pilots, 20 Farman observation pilots, and 9 Breguet-Michelin bombers.

Almost half of the pilots embarked for Salonika under the orders of Captain Robert Dangelzer. The plan was for them to reach Romania by air, but the weather got in the way and the airmen faced more than two weeks of continuous bad weather. On November 19, 1916, three Farman F.40 "Type Happe" attempted the crossing, piloted by Lt Bertrand de Fraguier, S/Lt Jules Mahieu, and a Romanian pilot, Lt Theodorescu; all of them brought their mechanic. As soon as they crossed the lines, they were confronted with a labyrinth of clouds and strong headwinds that forced them to return in a scattered order. On the way back, Mahieu's Farman flew over the town of Drama, where there was an enemy airfield: two German planes took off to attack it. The French crew faced up to the situation but their machine gun jammed. The soldier-mechanic Emile Maréchal, then seized the rifle and hit the first attacker who fell. Taking advantage of a short respite, he disengaged the machine gun and hit his second attacker! The two victories, confirmed by Allied troops on the ground, were duly recognized by the crew, who were even entitled to an article in Jacques Mortane's magazine, "*La Guerre Aérienne Illustrée*", in which the "coup du roi" of the machine gunner was celebrated. But this was the only success of this raid, because the third pilot, Lt. Theodorescu,

Above: General Henri Berthelot was sent to Romania at the head of the French military mission, which included an air contingent of 283 men, including 49 pilots. However, he could not avoid the invasion of two thirds of the country by the German, Austro-Hungarian and Bulgarian armies, but helped to hold a reduction in Moldavia in connection with the Russian army (Rol).

Above: This group of British pilots of the Royal Navy Air Service, stationed at Mudros, were sent by air to Romania during the summer of 1916. From left to right: Flight Lieutenant Arthur Frederick Foy Jacob, Flight Sub-Lieutenant Hervey, Flight Sub-Lieutenant Bush, Flight Commander Adams, Flight Sub-Lieutenant Reid, and an unknown pilot (Coll. *Cross and Cockade*)

was the last to return to Salonika around 4 pm and missed his landing, killing himself and his mechanic, Private Jules Manche. As the bad weather persisted, the commander of the Army of the East, General Sarrail, considered it preferable to stop these convoy flights, believing that the "Salonika-Romania link is a military sporting performance to be reserved only for trained pilots and choice aircraft". Paris agreed with him and on November 20, 1916 ordered the return of the pilots and equipment of the Romanian mission from Salonika, which were to reach their destination via Russia, through the port of Murmansk. The British, who were also present in Salonika, managed to send to Romania by air on October 25 and November 21 a total of 3 Nieuport 12 two-seater 110 hp Clerget, 3 Henri Farman HF 27 with 150 hp Canton-Unné engines as well as a Nieuport 11, but without mechanics or spare parts. Fortunately for the Romanians, a large part of the pilots and all the technical personnel of the French mission embarked for Romania from the port of Brest on October 10, 1916 and arrived at Arkhangelsk in Russia on the following 25, then traveling by train through Russia to Romania, which they reached in November, at the same time as all the senior officers destined to reorganize the Romanian army under the command of General Berthelot. Berthelot, who had just arrived, relieved Major de Malherbe of his command of the air mission on November 27 and appointed in his place a senior officer who was not an aviator, Major Rousse.

Rousse took command at the most critical moment, when enemy troops crossed the Danube and advanced into Wallachia, threatening the capital. The diagnosis he made in a report was very severe for the local Romanian airmen: "Of the Romanian personnel, I have the impression that none had yet stripped the peacetime airman to make him a man of war. The pilots, including the officers, exhausted their efforts in vain conversations and seemed, in general, accustomed to doing their will without concern for their military obligations. The observers practiced looking at the landscape while flying and did not report any useful information. As for the troops and the mechanics, their large number is still in inverse proportion to their technical value."

Immediately, he made French pilots and observers participate in war missions on the existing Romanian equipment, despite its very poor state of maintenance which caused the loss of two pilots, Lt de Maillé de la Tour Landry killed on his Farman on November 22, 1916, followed on November 27, 1916 by Lt Bertrand de Kergolay captured

Above: On the Mudros field, lined up in front of the airship, the three Nieuport 12s bearing RNAS serials n°8524 (French serial n° 1421, Flt Lt G.A. Cox's plane), 8514 (Flt SLt H.V. Reid) and 8525 (French serial n° 1422, Flt Lt A.F.F. Jacob). (Coll. *Cross and Cockade*)

Above: Flight Lieutenant G.A Cox departing for Romania on October 25, 1916, aboard his Nieuport 12 serial 8524 (and bearing French serial number 1421) (Coll. *Cross and Cockade*)

Rumpler G.II n°130/15 was shot down on 23 December 1916 at Ianca, on the Danube plain, by a Nieuport flown by Flight Lieutenant Arthur Frederik Foy Jacob, one of the members of the small British contingent reinforcing Romania. This was the first victory of the Entente air force in this campaign. The British contingent soon left the country, unlike the French aeronautical mission that was to form the backbone of the Romanian air force. (DR)

on his Nieuport in the enemy lines near Caracal due to a breakdown of his engine. At the end of November, as the battle of Bucharest began, the first "Franco-Romanian" squadrons were formed, equipped with Nieuport 11 fighters and Farman F.40 observation planes that had arrived from Russia. All of them had a two-headed command with a Romanian and a French officer, like the N.3 squadron created on November 25, 1916 for the defense of the entrenched camp of Bucharest, co-directed by Captain Nicolae Capsa and Captain Maurice Gond. It had only eight pilots (including two Romanians) and only six Nieuport fighters, of which only three were armed with a machine gun! The Nieuport 11 fighters were in poor condition, having suffered from their long journey, and were far from their optimum performance, which was already limited in normal conditions. They were unable to worry enemy aircraft, particularly bombers, which they were unable to reach in horizontal flight. The situation of the Farman F.40s carrying out observation and artillery adjustment missions was not much better; the first aircraft to leave the crates were Type 46s, whose De Dion engine gave serious setbacks. They were forced to fly low enough for artillery adjustment missions, and several of them came back riddled with bullets during the eight days of the battle of Bucharest – a French observer, S/Lt Georges Lebrun, came back wounded in the thigh during one of these dangerous missions.

The Romanian capital, attacked on 28 November, fell on 6 December 1916, but a few days earlier, Major Rousse had ordered the evacuation of his aircraft and, above all, of his equipment, including the precious Bessonneaux hangars, which took the road to the north and to Moldavia, where the remnants of the 1st, 2nd, and 3rd Romanian armies took refuge, while the 4th Army, which was still there, held the front in the mountains of Transylvania. It was during this confused period of retreat that the Romanian air force won its first two aerial victories on December 23, 1916 when a British pilot of the RNAS who had flown to Romania, Flight Lieutenant A.F.F. Jacobs, aboard a Nieuport 11, shot down one of the formidable Rumpler G.II (n°130/15) which was forced to land near the Ianca train station where its crew of 3 was captured – the aircraft was even taken back by train to the Russian lines. On the same day, a French crew of the N.3 squadron flying a two-seater Nieuport 12 of British origin, Warrant Officer James Texier with Lt Roger Lucy as observer, shot down a two-seater which landed in a field where it turned over with both wheels in the air, near the village of Rosiori south of Rimnicul-Sarat (victory confirmed by a later aerial reconnaissance).

3.4 – The fixation of the front

At the end of December 1916, a new defensive front line managed to be organized along the Carpathian mountains, the Sereth river which flows into the Danube, and on the mouth of the Danube itself. It was the Russian army that made it possible to re-establish the military situation by holding most of the new front. In the north, facing the Carpathians, the Russian 9th Army, which had held the southern part of the Russian front before Romania entered the war, extended its deployment area. Then, between the localities of Onesti and Adjud, the Romanian 2nd Army (former 4th Army) held the next segment of the front. The crucial sector of the confluence of the Sereth and Putna rivers is held by the Russian 4th Army of Ukrainian General Alexander Ragoza. The rest of the front, using the natural barrier of the Danube up to the Black Sea, was held by the Russian 6th Army.

The Romanian government only controlled the province of Moldavia and established its provisional capital in Iasi (Jassy in the French documents of the time). It was a rural and poor region with few resources, and it welcomed many refugees from all over Romania... As every year in this season, snow covered the region and temperatures dropped to about ten degrees below zero. But the winter of 1917 was particularly harsh for its inhabitants, especially since a terrible typhus epidemic was raging there.

It was in these very difficult conditions that General Berthelot's mission set to work to reorganize the Romanian army, equipping it in particular with machine guns and artillery pieces and above all reforming its chain of command. As far as the air force was concerned, Major Rousse left Romania at the beginning of 1917 to supervise railway logistics in Russia. He left his place at the head of the air mission to Lt-Col François de Vergnette de Lamotte, an old military pilot (military patent n°139 of July 25, 1912) who was given the task of continuing his predecessor's action aimed at organizing all the services of the new Romanian air force, which was going to have to support not only the Romanian army but also the Russian armies assigned to the Romanian front.

It was already necessary to organize the very long logistic chain from Russia, which ended in the city of Iasi where the Romanian aviation depot and its reassembly workshops were installed. The Russian railroads had only a small capacity and the Russian government would only cooperate with reluctance, not very happy to see deliveries of aircraft arriving in Arkhangelsk and escaping from its own air force, which itself lacked everything. Moreover, if the French pilots were to constitute the backbone of the new Romanian aviation (a total of 70 licensed pilots were to follow one another), it was necessary to train local pilots, observers, and mechanics. A pilot school was organized in Odessa under the command of Captain Paul Schneider. Two mechanics schools were opened in Iasi and Galati, as well as an observation and aerial photography school.

The logistical chain that had been set up made it possible to accumulate equipment: by 29 December 1916, 77 aircraft had been delivered to Romania, including 23 Nieuport fighters (only 7 were armed...), 36 Farman reconnaissance aircraft, and 18 Breguet V and Breguet-Michelin bombers that could not be used because they had not received their bomb launchers. On March 18, 1917, the total number of aircraft delivered had risen to 100 (22 Nieuport 11, 5 Nieuport 12, 48 Farman F.40, 3 Henri-Farman HF 27, 22

Above: What remains of the Romanian army as well as many civilians take refuge in Moldavia in dramatic conditions. It is a very poor region, overcrowded with refugees, and where a terrible typhus epidemic is raging (Photo Daniel Nemoz, Coll Passot-Seguin)

Above & Below: The Romanian troops and the Russian army (whose cavalry troops can be seen here) joined forces to resist there until the end of the war... The French pilots settled there under a polar cold and stayed with the inhabitants in very precarious conditions. (Photo Daniel Nemoz, Coll Passot-Seguin)

Breguet and Breguet-Michelin), but only 60 or so were in service, as 40 had already been destroyed, mainly due to the wear and tear of the airframes, which had been severely tested by the climate.

This delivered equipment allowed the creation of 8 squadrons divided into three aeronautical groups, each of them at the disposal of the three armies holding the center of the front. The first group, with its headquarters in Bacau, constituted the aviation of the 2nd Romanian army with the N.1 fighter squadron (on Nieuport) and the F.2 and F.6 observation squadrons (on Farman). The 2nd group, at the disposal of the 4th Russian army, was assigned the fighter squadron N.3 and the observation squadron F.4 which were all based at Tecuci – they reinforced the local Russian air force composed of the 4th, 14th, and 19th KAO as well as the 4th AAO. Finally, the 3rd group assigned to the 6th Russian Army included the F.5 and F.7 observation squadrons at Galati which reinforced the 4 Russian squadrons composed of the

Left Promoted to the rank of captain and appointed head of the N.1 fighter squadron, René Chambe (center) touches the propeller of one of the Nieuport 11s of his unit, stationed near Onesti in early 1917.

Facing Page, Above On February 14, 1917, S/Lt Charles Rivol-Tissot and Corporal Henri Manchoulas shot down a German aircraft in their lines and landed beside their victim. The Nieuport n°1413 of Corporal Manchoulas is seen here, on the left of the cockpit, surrounded by soldiers of the 4th Russian Army who witnessed the air combat (SHD)

Left View of the Nieuport 11 n°1405 of Capt. René Chambe at Onesti in early 1917. At the end of 1916, the aircraft was wearing the pennant of the French N12 squadron, which seems to have been removed here, but the aircraft may have been re-covered. Later in 1917, the order was given to display the Romanian colors on the aircraft, which in practice consisted of painting the white of the French roundels in yellow. (Coll Emmanuel de Vachon)

20th and 36th KAO as well as the 4th and 10th FAO. The BM.8 squadron on Breguet and Breguet-Michelin bombers remained in training for a long time in Iasi before being assigned to Galati under the direct orders of the general staff. Finally, let's mention the north of the front held by the 9th Russian army, which was not assigned any Romanian squadron but had 8 Russian squadrons, including one fighter squadron, the 9th FAO, led by the poruchik (lieutenant) Ivan Loïko, who was to end the war with 6 air victories.

Very quickly, the French equipment showed its insufficiencies: the delivered planes were outdated models having easily one year of delay compared to those in service on the French front. Romania was far from being a priority front for the French GQG... The Nieuport 11 (known as "13 meters" – their wing area) and Nieuport 21 (known as "15 meters") fighters had very limited power with their 90 hp engine, which did not allow them to join the best enemy aircraft in horizontal flight. The logbook of the N.3 squadron illustrates this perfectly by narrating their misadventures: "[On January 14, 1917] At 1:20 pm our pilots were at the table and some mechanics were hovering around the aircraft when they saw an aircraft coming slowly to Tecuci. This aircraft had a strange shape and already the mechanics thought they saw a SPAD. The plane was still descending and passing 20 m above the hangars it had cut its engine and was landing when we noticed that it was a Kraut, a smart

Facing Page, Below: The city of Bacau, flown over by a French Farman in early 1917. The dictator Ceaucescu having remodeled the historical city center of this city as well as many others, there is little left today of the streets and buildings appearing on the photo, except for the Saint Nicolae church which is on the right on the square in the center of the photo (Coll Passot-Seguin)

Farman F.40 with unknown serial number decorated with an angry Oran-Outang. It is an aircraft of the F.4 squadron stationed at Tecuci (DR)

Kraut who thought he was home! But it is necessary to believe that the Boche have an "old good God" who watches over them because at the moment when he was going to touch the ground we can clearly see the Boche observer who had probably seen the roundels of our aircraft gesticulating like a lost man. The pilot, realizing his blunder, turned the throttle back on and their "good old God" animal allowed his engine to start up again, and they took off again at the speed of a zebra. In the meantime our pilots had not wasted their time and at the cry of "un boche!" they had jumped from the table into their cuckoos. The pursuit began. Our pilots were about 800 meters behind, the Boche was gassing terribly and he was able to return to his den without being caught.

This did not prevent the French pilots, and especially those of Tecuci's N.3 squadron, from achieving several successes when they had the advantage of altitude. On February 12, 1917, Corporal Henri Manchoulas scored the squadron's second victory by downing an LVG south of Mărășești, 5 km into the enemy lines – the victory was certified after the wreckage of the enemy aircraft was photographed by a Farman. Two days later, on February 14, Manchoulas repeated the feat by coming across an enemy aircraft that he shot down in collaboration with S/Lt Charles Revol-Tissot. On February 23, 1917, it was Warrant Officer James Texier's turn to add a piece to his hunting record... Despite their equipment, the pilots of Captain Gond's N.3 won a collective total of 10 victories during the year 1917, which earned them a collective citation, the only squadron in Romania to be so distinguished.

Their observation colleagues were not much better off with the Farman F.40, an aircraft for which the crews reproached its vulnerability to attacks from the rear because

Above: This Hansa-Brandenburg C.I of the Austro-Hungarian Air Force, bearing the serial number 269.04, is produced by the UFAG company. Equipped with an Austro-Daimler engine of 200 hp, it is much more powerful than all the Romanian aircraft then in service on the front. It is seen here surrounded by airmen who invited some elegant ladies to its field in Transylvania (Coll B. Toschinger)

Nieuport 11 n°1413 of corporal Henri Manchoulas, Escadrille N.3, Tecuci early 1917. Manchoulas won 3 confirmed victories on the Romanian front.

Above: Hansa-Brandenburg C.I n°63.26, produced by the UFAG company, and equipped with a 160 hp Mercedes D.III engine. One notes the Schwarzlöse machine gun of 8 mm handled by the observer, statutory machine gun of the Austro-Hungarian army, but rather unsuitable for its use on the planes because of its weak rate of fire. (Coll B. Toschinger)

of the engine placed in the nacelle behind the pilot and the observer. They had to be escorted by fighters to be able to work safely, which was not always the case due to the lack of available fighters. Moreover, the De Dion engines that equipped the Model 46 had a limited reliability, which was not the case for the Renault-powered Model 42 delivered later. As for the bombing pilots on the Breguet V or Breguet-Michelin (see *Fana de l'aviation* n°614), the head of the French mission noted: «[They] gave us nothing but setbacks: poorly functioning engines, numerous breakdowns and very low performance. Their use has been almost nil and their reform is necessary in a short time.

3.5 – Rising to power and offensive actions - The battles of Mărăști and Mărășești

In April 1917 the work of the Berthelot mission to reorganize the Romanian army bore fruit and allowed the introduction of a new 1st Romanian army on the front, which came to be inserted between the 4th and 6th Russian armies. The new equipment delivered allowed the creation of four new squadrons, the F.9 observation, N.10 and N.11 fighter, as well as the C.12 on Caudron G.4 twin-engine aircraft whose strength was reduced. In July 1917, a total of 165 aircraft were delivered and 81 were in service (22 Nieuport 11, 8 Nieuport 21, 40 Farman, 7 Breguet and Breguet-Michelin, and 5 Caudron G.4) plus 19 others held in reserve or under repair, with 150 licensed pilots, 42 of whom were French, as well as 104 observers, 20 of whom were French.

Above: Nieuport 11 of Sergeantul Nicolae Mănescu, of N.1 Squadron. This aircraft was a reconditioned aircraft delivered to Romania (Coll Jack Herris)

Below: Breguet-Michelin BM 4 (n°256) just reassembled in the hangars of the Romanian General Aviation Reserve in Iasi, from where it will be given to the leader of escadrille BM.9, capitaine Armand Delas. These bombing planes were not very reliable and had a very low availability; they were used to attack the river navigation on the Danube (SHD)

The Marquis Is Chivalrous!

Above: Marquis Joseph Costé de Triquerville.

Above: Sergeantul Dumitru Bădulescu, Farman observer and the only Romanian aviation ace with 5 certified victories. (DR)

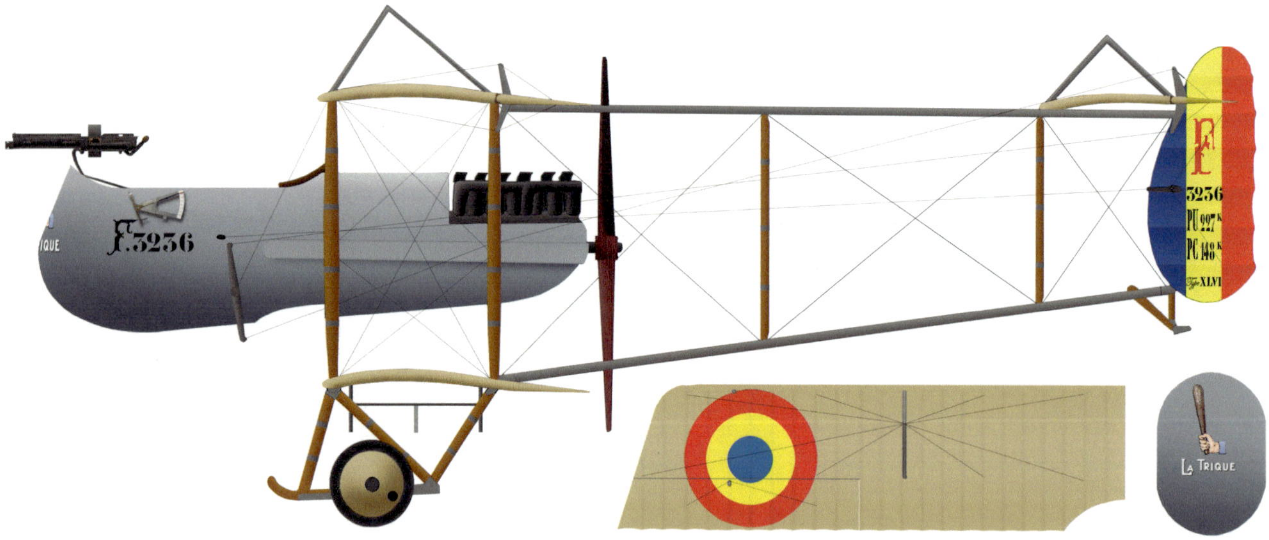

Farman F.40 n°3236, Escadrille F.2, Onesti, summer 1917.

Above: Farman F.40 n°3236 of escadrille F.2 at Onesti, seen in the spring of 1917. This is the personal aircraft of sergent Joseph Costé de Triquerville, who decorated it with a personal insignia, a hand holding a club, and gave it the nickname "La Trique" ("The club") in relation to his surname (DR).

Coming from an old French aristocratic family, the Marquis Joseph Costé de Triquerville was also mayor of the small town of Triquerville and owner of the stud farm of the same name. When the war broke out, he was a 34 year old father, who had been refused military service for weakness and bronchitis. He nevertheless wanted to serve and voluntarily enlisted in August 1914, finding himself a driver at the Saint-Cyr aviation center near Versailles where he learned to fly a plane "on the job" and became an instructor there in 1915 and 1916, obtaining the rank of corporal and then his military pilot's license in September 1916, thus regularizing a de facto situation. Two weeks later, he was assigned to the Romanian air mission, which he joined via Russia.

Promoted to the rank of sergeant, he was assigned to the F.6 observation squadron based in Onesti where he was given the Farman F.40 n°3236, which he christened "*La Trique*" in reference to his patronymic, decorating it with a fist firmly holding a club... On August 9, 1917, during the battle of Mărășești, he was ordered to carry out a long-distance reconnaissance mission by going to photograph the airfield of Târgu Secuiesc, 50 km inside the enemy lines, in Hungarian Transylvania. To protect him from any bad encounters, a Nieuport 11 fighter of the N.1 squadron piloted by Lieutenant Gheorghe Mihăilescu accompanied him. The outward journey is made without difficulties and his observer, Sergeant Dumitru Bădulescu, was able to take his pictures. But the return trip was much more eventful as they crossed paths with an enemy two-seater, an Oeffag C.II (No. 52.63) of the Austro-Hungarian air force, which surprises them and fires a burst from its front machine gun. Aircraft riddled but crew unharmed, the Farman was rescued by its guardian angel as Mihăilescu's Nieuport slipped behind the Oeffag and shot it down between the lines, where it made an emergency landing by breaking its landing gear.

From his forward position, Bădulescu saw the enemy aircraft catch fire as the crew appearred trapped in the fuselage. He exchanged a look with his pilot... Sergeant Costé de Triquerville understands. And he attempted a unique rescue in the history of the First World War: he landed next to his enemy, knowing that the operation was not at all obvious on the wooded hills of Transylvania, and moreover within range of the Austro-Hungarian and Romanian troops! No one dared to shoot at the Farman, which took off again immediately after leaving Dumitru Bădulescu on the ground. When he saw that the enemy pilot was dead, he managed to save the observer, whose legs were broken, from the blaze. He hoisted him on his back and carried him to the Romanian lines where the doctors managed to save his legs.

De Triquerville and Bădulescu went to visit their prisoner, Leutnant Francisc Xavier Schlarbaum, in the hospital in Iasi, whose enemy air force came to ask for news the next day by dropping a ballast message on the field of Onesti. The Franco-Romanian crew would answer them personally by throwing a message on the field of Târgu Secuiesc, to which the Austro-Hungarian airmen would answer by parachuting a box containing clothes and chocolate for Schlarbaum... and bottles of champagne for their enemies!

Rewarded with a citation in the army order by General Berthelot for this exploit, Sergeant Costé de Triquerville finished the war at the Chartres pilot school, then returned to his town hall and his horses after the armistice. Decorated with the Legion of Honor in 1949 for his breeding activity, he died in Paris in 1956. As for Dumitru Bădulescu, he ended the war with 5 homologated victories, all won as an observer, making him the only "ace" of Romanian aviation. Remaining in the army during the interwar period, he served again in the Romanian Air Force during World War II in a supervisory position and was imprisoned for two years by the communist regime during the Cold War. He died in 1978.

This change in manpower led to a new distribution of air resources on the front: the 2nd Romanian army and the 4th Russian army kept their air resources (respectively squadrons N.1, F.2, and F.6 for the first one, and N.3-F.4 for the second one) while the 1st Romanian army was assigned the fighter squadron N.10 plus the reconnaissance squadrons F.7 and F.9. The 6th Russian army, in addition to its own squadrons, received the reinforcement of the N.11 and the F.5. The BM.8 bombing squadron and the C.12 reconnaissance squadron were placed under the direct orders of the general staff. Lt-Col de Vergnette, widely supported by General Berthelot, asked Paris for more modern aircraft, SPAD fighters for the fighter, or at least Nieuport 17s, and Sopwith two-seaters for observation, all in order to obtain at least technical parity with the opposing air force, which was not very aggressive because it was made up of Austro-Hungarian squadrons. The GQG in France granted his request in May, but due to logistical difficulties, it was not until September that they were actually delivered to the Romanian air force... That is to say, after the great battles of the summer of 1917.

The first of these, known as the Battle of Mărăști, took place from July 22 to August 1, 1917. The Romanian general staff ordered the 2nd Romanian Army to attack in the Carpathian Mountains, facing the troops of the 1st Austro-Hungarian Army, which they managed to push aside. The Romanian artillery, well trained and supervised by French officers, were able to do a very efficient job with the help of the adjustments made by the Farman planes of the F.2 and F.6 squadrons, which redoubled their activity, being protected by the few fighter planes of the N.1. In spite of the low number of troops, the use of aerial means was optimal as the Austro-Hungarian general Friedrich von Gerok noted: "The Romanian artillery is very active in the attack. Thanks to the collaboration of its air forces, with a strong bombardment, it executed all the breaches just before the attack. Our artillery, given the excessive activity of the enemy air force, is no longer able to silence the enemy batteries.

The front was thus broken over a length of thirty kilometers and over a depth of twenty kilometers, at the cost of the loss of 5,000 Romanian soldiers against enemy losses twice as high. Unfortunately, this important tactical success could not be exploited by the 4th Russian army, which was supposed to attack in the plain: the Russian government forbade it to go on the offensive and took elements to assign them to other fronts.

Indeed, two weeks earlier, the last offensive of the Russian army, led by the head of the government Alexander Kerensky, who took the reins of the country after the abdication of the Tsar, took place in Ukraine. His offensive, after some initial successes, quickly turned into a disaster and led to serious

Above: Nieuport fighter of Sergent Henri Manchoulas on patrol along the Putna River, May 23, 1917, photographed by a Farman it was escorting. Moments later, attacked by an enemy aircraft, Manchoulas was wounded in the head and had his machine gun destroyed, but remained to protect the Farman. He was awarded the military medal for this feat (DR).

Below: Pilots from French-Romanian escadrille N.3 around March 1917. From left to right: sergent Henri Manchoulas (3 victories), adjudant Charles Revol-Tissot (2 victories, +1 won in France), capitaine Maurice Gond (2 v.), adjudant James Texier (3 v., killed on August 16, 1917), caporal Maurice Théron (1 v.), sergent François Terry (SHD)

Left: Capitaine Maurice Gond, commanding officer of escadrille N.3, in front of his Nieuport decorated with an archer (SHD)

Facing Page, Above: Caporal William Billon du Plan (2nd left) of escadrille F.7 squadron poses with French and Romanian squadron mates in front of Farman F.40 n°3233 (DR)

Facing Page, Below: The Farman F.40 n°3197 of escadrille F.5 is decorated with a legendary figure of the Romanian nation, Michael the Brave (Mihai Viteazul), Prince of Wallachia and hero of the fight against the Turks at the end of the 16th century (DR)

Below: Adjudant Charles Revol-Tissot holding a Romanian pilot by the shoulder, in front of a Nieuport 11 which could be the latter's. In fact, in a correspondence written in the 1960s, Revol-Tissot indicated that his personal Nieuport was decorated with a red five-pointed star (DR).

F.3197

Above: Romanian soldiers (including a naval officer on the left and a pilot on the right) enjoy themselves in front of a portrait in Borcesti in October 1917. They probably belong to escadrilles N.1 and F.2 (Photo Daniel Nemoz, Coll. Passot-Seguin)

Left: This Romanian pilot is staying with peasants of Borcesti during the autumn of 1917 (Photo Daniel Nemoz, Coll Passot-Seguin)

Above: An airfield of a Fliegerkompagnie of the Austro-Hungarian Air Force in Transylvania during the summer of 1917 (Coll B. Toschinger via Boris Ciglic)

Above: The Hansa-Brandenburg KD, designed by Ernst Heinkel, was a rather ugly fighter nicknamed "the spider" because of its crossed struts, and had the flying qualities of an iron. Some of these aircraft, such as this one with the number 28.47, were in service on the Romanian front (Coll B. Toschinger)

Nieuport 11 n°1413 of Adjudant James Texier (Escadrille N.3, 3 victories), the only French pilot of the Romanian mission killed in air combat.

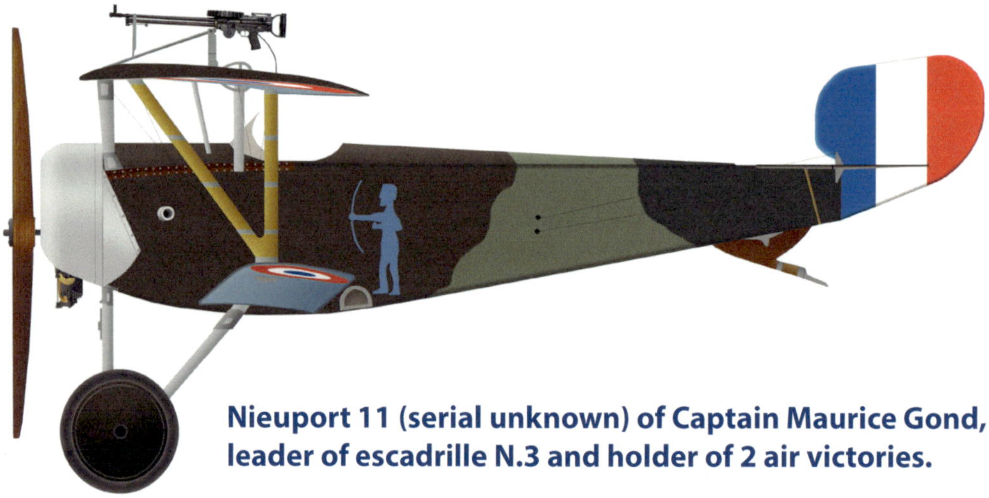

Nieuport 11 (serial unknown) of Captain Maurice Gond, leader of escadrille N.3 and holder of 2 air victories.

Breguet Michelin BM 4 n°256 of capitaine Armand Delas, leader of escadrille BM.8, Galati, summer 1917.

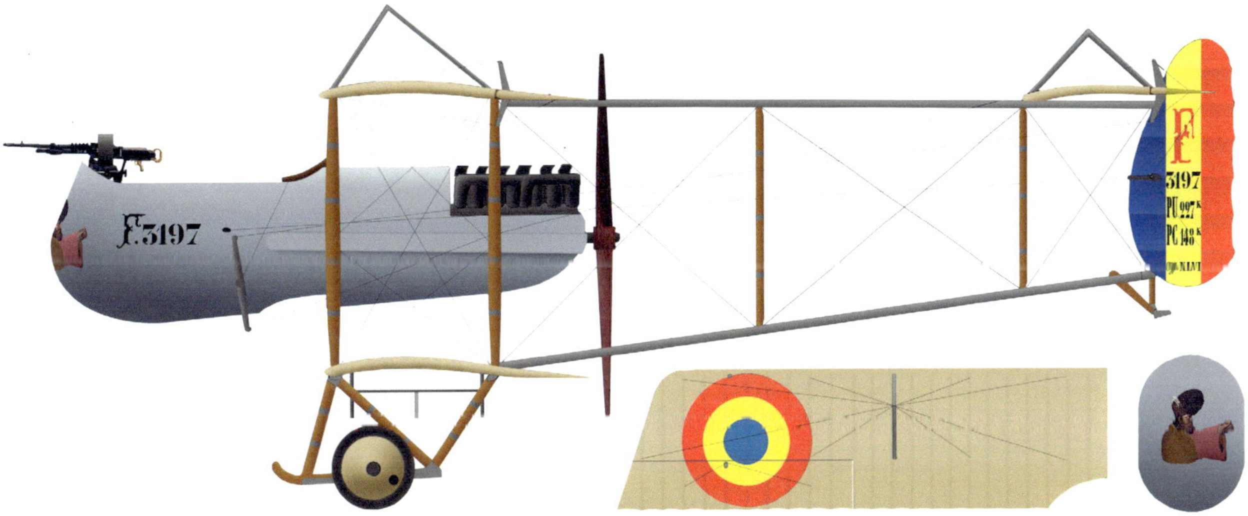

Farman F.40 n°3197, Escadrille F.7, Tecuci, summer 1917.

Hansa-Brandenburg KD n°28.47 of the Jagd-Flieger Detachment Nikitsch, Kezdi-Vasarhely (currently Târgu Secuiesc), summer 1917.

Albatros D.III n°53.24 of Hptm Nikitsch, Jadg-Flieger Detachment Nikitsch, Romanian Front 1917.

Above: Albatros D.III n°53.24 of Hauptmann Nikitsch, commander of the Jagd-Flieger Detachment bearing his name, and gathering all the fighters of the surrounding Austro-Hungarian squadrons, forming a kind of marching squadron of fighters. Nikitsch, who finished the war with 6 victories, won 5 on the Romanian front (including one on board this aircraft) in the middle of a sector held by the Romanian air force, but none of his claims in Romania can be found in the French or Romanian archives (Coll B. Toschinger)

mutinies in the army where the soldiers, exhausted by three years of war and worked by Bolshevik propaganda, left their units en masse to return home, sometimes executing their officers... The troubles spread throughout the Russian army from the Baltic to the Black Sea and the 4th Russian army of the Romanian front commanded by General Alexander Ragoza was no exception, although it managed to maintain its cohesion and at least hold its sector.

The German 9th Army under General August von Mackensen, which occupied the Danube plain and had to intervene during the battle of Mărăşti to fly to the aid of the Austro-Hungarian army, intended to take advantage of the Russian weakness and as early as August 6, 1917, launched an offensive in the plain along the Sereth River, precisely on the sector of the 4th Russian Army with its relative fighting spirit. The logbook of the N.3 squadron, which multiplied the missions on the battlefield, testified to a tragicomic incident that occurred on August 8: "At 4 pm, Théron leaves to accompany Farman. On the way back, wanting to show friendship to the tenant of a Russian sausage, he approaches the basket and makes many friendly signs to him. But the "saucisman", who had been attacked in the morning by a Boche, was on the defensive and did not understand Theron's urbanity. He grabbed his machine gun and sent a volley of bullets at our pilot, 14 of which arrived in the aircraft. Theron still didn't understand and figured that this was another consequence of the Revolution." Von Mackensen hoped to achieve a decisive breakthrough and liquidate the Romanian front... His attack met with initial success by routing the 4th Russian army. But this was without the resistance of the Romanian troops who fought to keep what remained of their country: The 2nd Romanian Army in the west and the 1st Romanian Army in the east both extended their front lines to replace the Russian troops and succeeded in thwarting the German attack after more than four weeks of intense battle that ended on September 8 and that history remembers as the battle of Mărăşeşti, a village that became the Romanian Verdun at the cost of 27,000 Romanian soldiers and 25,000 Russians, who put 65,000 German soldiers out of action. It was also during this battle that the only French pilot killed in aerial combat fell, Warrant Officer James Texier holder of 3 certified victories. On August 15, 1917, while on patrol with Warrant Officer Manchoulas, he encountered an enemy two-seater on which he was attacking. The enemy machine-gunner was skilful and hit him with a bullet in the belly, forcing him to break off the fight and to land in a field from where he was taken to the hospital of Tiganesti. The squadron's diary notes about him on August 16: "Nice weather. A day of sorrow and mourning for the squadron, our poor little Texier who was wounded in the stomach during the previous day's combat and who was taken to the hospital in Tiganesti died there in the evening of an internal hemorrhage. All of us, except Theron who was in Iasi, spent some time at his bedside and were able to say goodbye. In the evening at 10 o'clock, when Colonel de Vergnette came to give him his military medal, Texier died gently like a little bird. The King of Romania himself came the next day to pin on his deathbed the medal of military virtue (Virtutea Militară) in gold, before he was buried in the cemetery of Tiganesti.

Above: Farman F.40 with Romanian cockades on its nose in a snowy field... A symbol of all Romanian aviation at the beginning of year 1918 : left on her own with the departure of the French. (Coll. Emmanuel de Vachon)

3.6 – An armistice forced by the Russian defection

What the German army could not obtain by force, the Russian empire would obtain by its defeat. The Russian army in Romania disintegrated in the autumn of 1917 as the country was shaken by the wave of desertions following the failure of the Kerensky offensive in July, and the attempted coup d'état by General Kornilov in September. The seizure of power by the Bolsheviks on November 7, 1917, led to the collapse of what remained of the army's cohesion and officially brought the former Russian Empire out of the war. Romania saw not only the departure of the Russian troops on its territory, but also the loss of its only supply route to the Western countries... Moreover, German troops took advantage of the chaos and advanced into Ukraine without encountering any resistance, threatening the territory controlled by Romania with a complete encirclement. Also, despite the exhortations of General Berthelot, who was determined to organize the resistance in Moldavia, the Romanian government decided to throw in the towel and signed the Focsani armistice with the Central Empires on December 9, 1917.

This decision led to the repatriation of the entire French military mission in Romania, including the air mission, which left the country in February 1918, leaving 78 aircraft in flying condition, including 18 Nieuport 110 horsepower aircraft (type 17 or 23), 16 Nieuport 80 horsepower aircraft (type 11 or 21), 26 Farman F.40s, 1 Henri-Farman HF 27, 14 Sopwith 1A2s, 2 Caudron G.4s and 1 Morane-Saulnier.

The return to France of the French airmen was not without difficulties, because they took the same route as the outward journey through Russia, which was now in the middle of a civil war. Captain Maurice Gond, who left the front shortly before the Focsani armistice, found himself in St. Petersburg in December 1917 and was robbed of all his savings by the Bolshevik customs officers during his passage to Finland – a tidy sum of 11,000 Romanian lei, including 2,000 in bonuses for his two air victories. On January 16, 1918, Corporal Claude Veau, a mechanic on his way to the Odessa flight school, which was later evacuated, saw his train stopped and looted by a gang of Russian soldiers led by an officer, who stole his wallet, his weapon, and his provisions. The Romanian officers on the train also suffered the same fate...

In one year of aerial combat, the 70 French pilots formed the backbone of the Romanian air force, with 12 squadrons, including 4 fighter, 7 reconnaissance, and 1 bombing squadron. They won 20 certified victories, plus about thirty others claimed by the Romanian pilots, of which about 120 were certified. Despite the technical inferiority, this small aviation fulfilled its mission by ensuring the surveillance of the front and the indispensable artillery adjustment that allowed the success of the battles of Mărăști and Mărășești, with very light losses since there were only three French

Above, Above Right, & Right: Farman French soldiers attached to escadrilles N.1 and F.2 are walking in the snowy village of Borcesti, in January 1918. They will soon pack up, forced by the armistice signed by Romania. (Photo Daniel Nemoz, Coll. Passot-Seguin)

pilots lost in combat (2 killed, 1 taken prisoner) and two others in accidents.

3.7 – Neither war nor peace

If the French had left the country and a cease-fire is effective with the German and Austro-Hungarian troops, the air force as well as the Romanian army remained on the war footing, against a new enemy: all the territory under the direct control of the Romanian government was occupied by Russian troops left to their own devices and engaged in banditry to survive. Romanian planes continued to fly over Moldova to monitor the departure of these troublesome guests. Although there were no longer any French airmen to fly them, there was still a discreet senior staff under the command of Major Jean Dufaure de Rochefort, a very experienced air observer and veteran of the 1915 Serbian campaign, who was appointed commander of the air force assigned to operations in Bessarabia.

For it was precisely in this province that all the action of the Romanian aviators will be concentrated during the whole year 1918. Eastern part of the historical Moldavia located between the Pruth and Dniestr rivers, it belonged to the Russian Empire but the majority of its population was composed of Romanians. In the chaos that gripped the former empire of the Tsars after the Bolsheviks took power in Moscow and Petrograd, several local assemblies were formed in the four corners of the empire and managed their provinces outside the central power. The parliament that was formed in Bessarabia, the Sfatul Țării, declared the autonomy of the province and then on January 24, 1918, declared its independence under the name of Moldavian Democratic Republic. The young state, threatened by the Russian armies debanded, white troops or red troops who plunder the country, called for help from Romania, which intervened with its army and air force to drive them out, under the benevolent neutrality of the German army, which advanced into Ukraine and installed a local dictator, the Hetman (Cossack leader) Pavlo Skoropadsky. The campaign ended on April 9, 1918, when the council of the Moldavian Democratic Republic voted to merge the province with Romania.

It was also at this time that, following the armistice with the Central Empires, the peace treaty talks were concluded in the form of the Treaty of Bucharest, signed on 7 May 1918. Romania, whose negotiators were able to take advantage of the differences between Germany, Austria-Hungary, and Bulgaria, did rather well because its territorial losses (the southern part of the Drobudja to Bulgaria, the rectification of the border in Transylvania to Hungary) were compensated by the gain of Bessarabia. However, its economy was cut off by Germany and a military force of the Central Empires remained to occupy the conquered country while the Romanian army remained mobilized in Moldavia and Bessarabia. The Romanian population continued to suffer severely from the requisitions and this treaty, which was never ratified by the Romanian parliament, was only a transitional situation because as the months passed it became

Above: Romanian pilots of escadrille S.5 n in front of a Sopwith 1A2, on the Bessarabian front in February 1918. The aircraft bears the motto "Hogne qui voura" (Growls who will), the personal motto of the family of capitaine Augustin de Mailly et de Nesle, the second on the left in the picture. A few days later, this officer will have to go back to France, like all the pilots of the Romanian mission, leaving their Romanian comrades to their fate (DR)

clear in the summer of 1918 that Germany was going to lose the war...

Close contacts were discreetly maintained with France and on October 22, 1918, a Breguet 14 piloted by Lieutenant Louis Noël took off from Salonika and landed in Iasi, taking with it the Romanian minister Victor Antonescu, who informed the Romanian government of the prospects of defeat of the Central Empires. All the minorities of the old Habsburg Empire are agitated and the Romanians of Transylvania organized themselves in a national committee. While Austria-Hungary capitulated on November 4, 1918, the Romanian government decided to re-enter the war on November 7, 1918 and remobilized its army, which crossed the Transylvanian passes and occupied this region without any real fighting, while the Austro-Hungarian and German troops evacuated the country.

The Romanian air force resumed service during this blitzkrieg, but only to the extent of its very limited resources, as it had spent more than a year cut off from all supplies. Knowing that on the French front, planes are consumables that last only four months in the front line, one can assume that the available planes can be counted on the fingers of one hand! Only one aeronautical feat was to be noted when on November 24, 1918 Lt Vasile Niculescu taking Capt. Victor

Precup as an observer takes off from Bacău on a Farman F.40 out of breath, whose ceiling is limited to 2,500 m and does not allow him to fly over the highest mountains he has to cross in the mist... He nevertheless manages to pass and land in Blaj to distribute government documents addressed to the national council of Romanians in Transylvania.

Although the Rethondes armistice of November 11, 1918 officially put an end to the First World War, the fighting did not stop in Eastern Europe because of the many upheavals linked to the Russian revolution. The former empire of the Tsars changed its flag but kept its imperialist reflexes since Lenin ordered the Romanian government to return Bessarabia and to advance its troops... The French armée d'orient sent troops to secure the Bessarabian border, including air units. Their story was told in the book "*Oriental Adventures of the French Air Force 1914-1919*". Romanian troops were also standing in the north of Moldavia with a few aircraft, which would conduct some skirmishes with enemy aircraft on the north bank of the Dniester. On April 8, 1919, a Nieuport 23 piloted by S/Lt Ion Dragomirescu claimed an aircraft at Mohyliv-Podilsky. Then, on April 19, 1919, a Romanian two-seater (probably a Sopwith 1A2) piloted by Lt Nicolae Drosso with Lt Aurel Constantinescu as observer, also claimed a two-seater Anatra that fell on the eastern bank of the Dniester. All of them are described as Bolshevik aircraft, but one should not swear by anything, because at that moment of the civil war ravaging the former Russian Empire, it is the Ukrainian Republic that still controlled the main part of the northern bank of the Dniester river!

Moreover, in Hungary, a communist government led by Bela Kun took power on March 21, 1919 and intended to challenge the borders resulting from the armistice: a three fronts war would begin with French and Czech attacking from the North, Romanian from the East, and French from the South. All those operations will be treated in chapter 4.

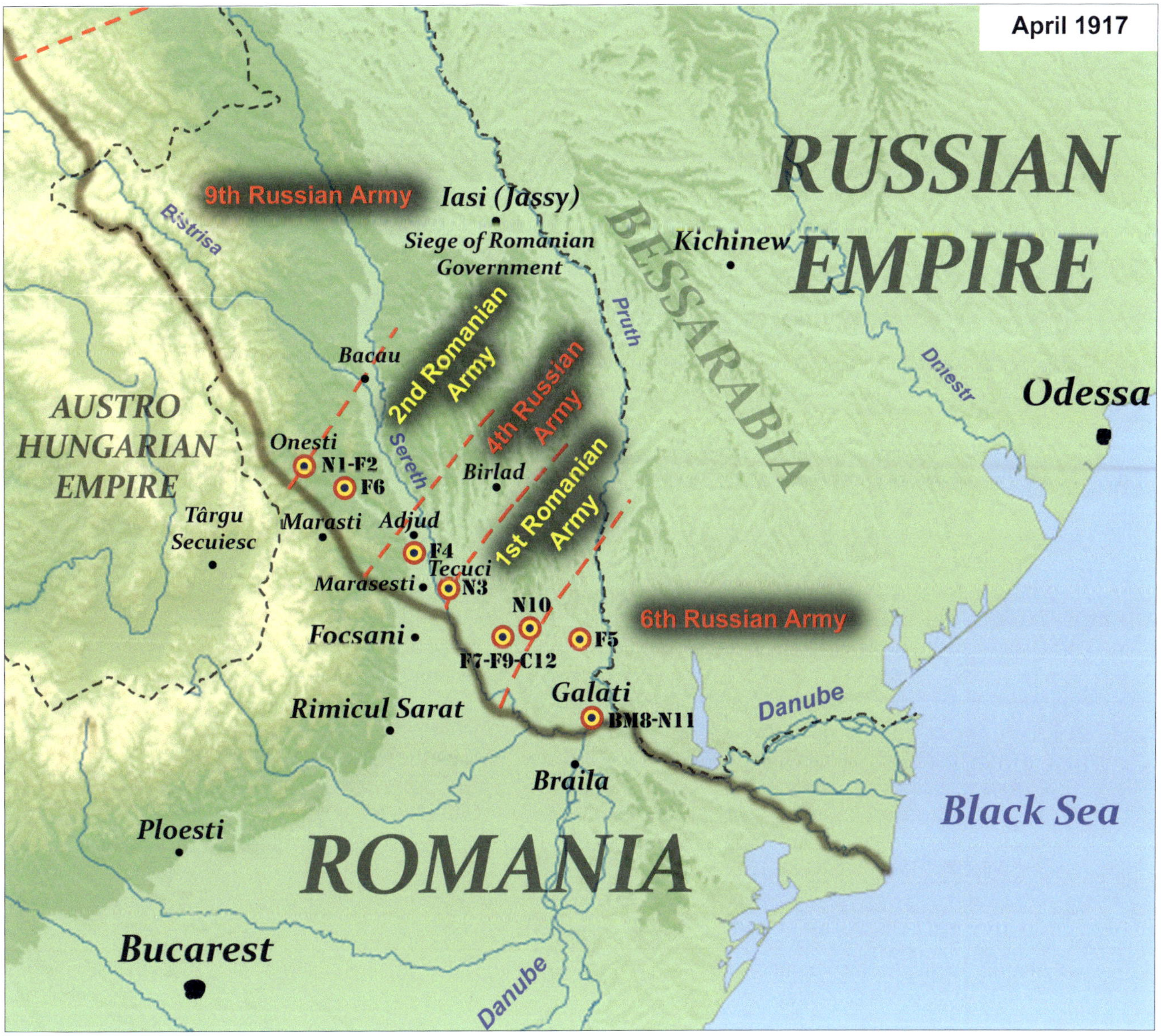

April 1917
RUSSIAN EMPIRE
9th Russian Army
Iasi (Jassy)
Siege of Romanian Government
Kichinew
BESSARABIA
Bistrisa
Pruth
Dniestr
Odessa
2nd Romanian Army
4th Russian Army
1st Romanian Army
Bacau
AUSTRO HUNGARIAN EMPIRE
Onesti
N1-F2
F6
Sereth
Birlad
Târgu Secuiesc
Marasti
Adjud
F4
Tecuci
Marasesti
N3
N10
6th Russian Army
Focsani
F7-F9-C12
F5
Galati
Rimicul Sarat
BM8-N11
Danube
Braila
Black Sea
Ploesti
ROMANIA
Bucarest
Danube

Victory Claims on Romanian Front by French Pilots or

#	Date	Rank	Pilot or Gunner		Unit	Plane Piloted	Plane Claimed	Place
1	12/23/16	Adj	Texier, James (Pilot)	1	N.3	Nieuport 12 (n°3978 british)	AEG	N. Rosiori (Rimnicu-Sarat)
		Lt	Lucy, Roger (Observer)		F.2			
2	2/12/17	Cal	Manchoulas, Henri	1	N.3	Nieuport 11	LVG	Marashesti
3	2/14/17	S/Lt	Revol-Tissot, Charles	2	N.3	Nieuport 11	Albatros C	Diochesti (SW Birlad)
		Cal	Manchoulas, Henri	2	N.3	Nieuport 11		
4	2/23/17	Adj	Texier, James	2	N.3	Nieuport 11	LVG	Foscani
-	4/2/17	Cne	Bléry, Paul (Pilot)	-	N.10	Nieuport 12	LVG	Isaccea, near Galati
		Adj	Tresch, Henri (Observer)					
-	4/2/17	Sgt	Bentzmann (-de), Christian	-	N.10	Nieuport 11	Seaplane	Canal Macin, near Reni (Danube)
5	4/13/17	S/Lt	Revol-Tissot, Charles	3	N.3	Nieuport 11	Albatros C	E. Focsani - 3 Km O. Parenpani
6	4/16/17	Cal	Théron, Maurice	1	N.3	Nieuport 11	Fokker	Bilieshti
-	1-May-17	Sgt	Blanchet, André (Pilot)	-	BM.8	BM n°252	LVG	?
		Cal	Guimon Louis (Observer)					
-	23-May-17	Adj	Bentzmann (-de), Christian	-	N.10	Nieuport 11 n°1509	Aviatik C	Braila
-	23-May-17	Adj	Bentzmann (-de), Christian	-	N.10	Nieuport 11	LVG	Braila
		Adj	Brun, Georges	-	N.10	Nieuport 11		
7	31-May-17	Lt	Schneider, Paul	3	N.1	Nieuport 11	Austrian plane	Soveja
		S/Lt	Brullard, Louis	-	N.1	Nieuport 11		
8	25-Jun-17	Sgt	Lamproux, Donatien	1	N.1	Nieuport 11	Airplane	Magura Casin
9	25-Jun-17	Sgt	Manchoulas, Henri	3	N.3	Nieuport 11	Airplane	Chiuslea ((Padursa / Negra)
-	30-Jun-17	Lt	Flers (-de), Amédée	1	N.10	Nieuport 21 n°1951	Two-seater	Independenta (25 km from Calarasi)
10	8-Jul-17	Adj	Alexandre, Pierre (Pilot)	1	C.12	Caudron G.4	Two-seater	Faurei (village)
11	20-Jul-17	Cne	Gond, Maurice	1	N 3	Nieuport 21 n°1909	Biplace	Ciusela

Observers 1916–1917

Source	Confirmation	Notes
Log Book escadrille N.3	Citation OA 15-02-1917 - JORF 17-07-1917 p.5535	Enemy lines - Crash confirmed by other pilots from escadrille N.3
and Romanian communiqué	Citation OA 15-02-1917 - JORF 17-07-1917 p.5535	
Log Book escadrille N.3	Citation OA (MM) 24-07-1917 - JORF 26-07-1917 p.5816	
Log Book escadrille N.3	Citation OA 25-05-1917 - JORF 23-06-1918 p.5407	FA 41 - LdR Gottlieb Kienle (Observer) KIA in Tecuci
	Citation OA (MM) 24-07-1917 - JORF 26-07-1917 p.5816	
Log Book escadrille N.3	Citation OA (MM) 09-10-1917 - JORF 11-10-1917 p.8026	A Farman from escadrille F.4 takes a photo of the crashed enemy plane
Romanian communiqué	None.	
Romanian communiqué	Citation OA 20-08-1917 - JORF 24-12-1917 p.10530 "forced to alight" (NOT DESTROYED)	Forced to alight
Log Book escadrille N.3	Confirmed by general citation escadrille N.3 (JORF 11-02-1918 p.1492)	18h00. Russian troops witness victory - A Farman from escadrille F.4 takes a photo of the crashed enemy plane
Log Book escadrille N.3	Confirmed by general citation escadrille N.3 (JORF 11-02-1918 p.1492)	
Romanian communiqué	None.	BM damaged by the attack of the LVG he claimed.
Romanian communiqué	Citation OA 10-08-1917 - JORF 24-12-1917 p.10530 "enemy put to flight" (NOT DESTROYED)	
Romanian communiqué	Citation OA 10-08-1917 - JORF 24-12-1917 p.10530 "enemy put to flight" (NOT DESTROYED)	17h45.
Romanian communiqué	Citation OA 10-08-1917 - JORF 24-12-1917 p.10530 "enemy put to flight" (NOT DESTROYED)	
Romanian communiqué	Citation OA 05-06-1917 - JORF 07-11-1917 p.8880	
Romanian communiqué	No confirmation for Brullard	
Romanian communiqué	Citation OA 01-07-1917 - JORF 02-12-1917 p.9729	9h10. Flik 1 - Hansa-Brandenburg C.I n°67.13 Flik lost near Kimpolung (Pilot Zgsf Poschwitz and Obs. LdR Schonowski, MIAs)
Log Book escadrille N.3	Citation OA 01-07-1917 - JORF 02-12-1917 p.9729	FA 242w - Pilot WIA, Observer LdR Theodor Schultheis KIA.
Romanian communiqué	None.	
Romanian communiqué	Citation OA (MM) 23-09-1917 - JORF 25-09-1917 p.7592	Enemy crew reputed killed.
Log Book escadrille N.3	Citation OA (LH) 28-08-1917 - JORF 30-08-1917 p.6874	

Victory Claims on Romanian Front by French Pilots or

#	Date	Rank	Pilot or Gunner		Unit	Plane Piloted	Plane Claimed	Place
12	7-Aug-17	Adj Plutonier	Texier, James Popescu, Marin - ROMANIAN	3 2	N.3 N.11	Nieuport 11	Fokker E.III	Faurei, Putna valley
13	8-Aug-17	Adj S/Lt	Bonneton, Fernand (Pilot) Naderval (Observer)	1 1	F.7 C.12	Caudron G.4 n°1814	Fighter	N. Marasheshti
14	13-Aug-17	S/Lt	Flers (-de), Amédée	2	N 10	Nieuport 21	Airplane	Franshtal forest (Bukhovina)
		Adj	Brun, Georges	1	N 10	Nieuport 11		
15	16-Aug-17	Cne	Gond, Maurice	2	N 3	Nieuport 21	Two-seater	Palerne (3 km NW Birlad)
16	21-Aug-17	Adj	Bruyère, Max	1	N 1	Nieuport 11	Hansa-Brandenburg C.I	Harja (15 km SO Ocna)
17	8-Sep-17	Lt	Brullard, Louis	1	N 1	Nieuport 11	Airplane	Kherzha
-	8-Sep-17	Cne	Micheletti, Charles	-	N 1	Nieuport 11	Airplane	Magura Casinuliu
18	9-Sep-17	Lt	Brullard, Louis	2	N 1	Nieuport 11	Fighter KD	
19	20-Sep-17	Cne	Micheletti, Charles	1	N 1	Nieuport 11	Biplace	Mogura Casinuliu
20	30-Oct-17	Lt	Bonneton, Fernand	2	N 1 (?)	Nieuport	Hansa-Brand-enburg C.I	Magura / Tirgul Ocna
		Cne	Micheletti, Charles	2	N 1	Nieuport		
-	11-Nov-17	Cne	Bléry, Paul	-	N 10	Nieuport	Avion	Sereth river, direction Klimovitz
		Plutonier	Paucker, Carol - ROMANIAN	1	N 10	Nieuport		

French Losses on the Romanian Front 1916–1917

Date		Pilot (P) or Observer (O), Fate	Unit	Plane Piloted
11/22/16	Lt	Maillé de la Tour Landry (-de) (KIA)	French Mission	Farman F.40
11/27/16	Lt	Kergorlay (-de), Bertrand (POW)	French Mission	Nieuport
?/11/1916	S/Lt	Lebrun, Georges Louis (Observer) (WIA)	French Mission	
12/11/16	Cal	Lemonnier, Jean (KIAcc)	French Mission	Nieuport 21
7/24/17	Cne	Chambe, René (WIA)	N 1	Nieuport 11
8/15/17	Adj	Texier, James (WIA-DOW August 16, 1917)	N 3	Nieuport 11
11/13/17	MdL	Chandou, Marcel (KIAcc)	French Mission	Farman F.40

Observers 1916–1917

Source	Confirmation	Notes
Log Book escadrille N.3	Citation OA (MM) 09-10-1917 - JORF 11-10-1917 p.8026 "3 victories"	20h00
Romanian communiqué	Citation OA 10-08-1917 - JORF 24-12-1917 p.10530	FA 42 - Halberstadt D.II lost in sector.
Romanian communiqué	Citation OA 01-09-1917 "shot one plane"	
	Citation OA 10-08-1917 - JORF 24-12-1917 p.10530	
Log Book escadrille N.3	Citation OA (LH) 28-08-1917 - JORF 30-08-1917 p.6874	FA 20 - One two-seater lost in the same sector.
Romanian communiqué	Citation OA 28-08-1919 "Shot down enemy planes"	Flik 31 - Hansa-Brandenburg C.I n°67.40 captured (Hptm Aurel Schiavon and Lt Josef Schuff escaped in their lines)
Romanian communiqué	Citation OA 16-09-1917 - JORF 06-01-1918 p.281	Enemy plane confirmed shot down in enemy lines by a Farman from escadrille F.7
Romanian communiqué	None.	FA 42 - Plane lost in sector, Ltn Paul Guthoff KIA.
	Citation OA 16-09-1917 - JORF 06-01-1918 p.281	
Romanian communiqué	Citation OA 16-09-1917 - JORF 06-01-1918 p.281 (Victory dated September 21)	Flik 33 - Hansa-Brandenburg C.I n°269.34 destroyed on landing, pilot WIA.
Romanian communiqué	Citation OA	
	Citation OA	
Romanian communiqué	None.	

Cause	Place	Notes
Flak (most likely)	Bucarest	
Flak (most likely)	Caracal	During a mission on the Danube river
Flak (most likely)		Wounded in the left thigh in November 1916, no exact date known
Accident	Iassi	
Air combat		
Air combat	Tecuci	Wounded by the rear firing of a two-seater with a bullet in the stomach
Accident	Iassi	Candidate pilot killed during a training flight

Chapter 4: Against Red Hungary and Russia, French Missions in Czechoslovakia and Poland (1919)

Above: Bela Kun, ruler of communist Hungary in 1919 during 133 days from March 21 to August 6, 1919.

4.1- Hungary takes up arms (March 21, 1919)

The Bolshevik revolution came to power in a country other than Russia: Hungary, where a Bolshevik-inspired "republic of councils" held the reins of the country for 133 days, between March and August 1919, before being swept away by an international military coalition.

Since the compromise of 1867, Hungary had been part of the Austro-Hungarian Empire, whose head of state was the Emperor of Austria and King of Hungary, and could thus rely on the Magyar aristocracy to run its vast multi-ethnic empire. The kingdom of Hungary, or Transleithania, included a number of minorities such as Croats, Serbs, Romanians, Jews, and some Germans.

In 1918, this crumbling empire, exhausted by four years of war, was on the verge of exploding: famine was rampant in Vienna as well as in Budapest; the young emperor Charles I tried to negotiate a separate peace with the allies and it was more or less by force that his powerful German neighbor forced him to remain in the conflict. When the Italian troops broke through the front at Vittorio-Venetto on late October 1918, the whole empire fell apart. The proclamations of emancipation of the Czechs, Slovaks, Poles, Ukrainians, Bucovinian and Transylvanian Romanians, Serbs, Croats, and Slovenes de facto detached vast territories from the "double monarchy": an international commission, the Lord Commission, began in the spring of 1919 to draw borders between these territories that peace treaties would later formalize.

The Kingdom of Hungary was transformed into a republic led by the government of Count Mihály Károlyi. He was not helped by the intransigence of the Allies, who imposed cruel territorial cuts that went beyond the ethnic borders and put many Hungarians under the sovereignty of new

Above: The communist government of Bela Kun has at its disposal aircraft from the former Austro-Hungarian aviation, such as this Hansa-Brandenburg C.I marked with the red star.

states. On March 20, 1919, Lieutenant-Colonel Vix, a senior officer of the French army stationed in the south of Hungary and stemming from General Franchet d'Esperey's *Armée d'Orient*, handed over a note to the Hungarian government demanding the cession of new territories. The country was greatly reduced to the benefit of the new Yugoslavia (Banat region), of the new Czechoslovakia (through Slovakia, part of which was populated mainly by Hungarians) and above all of Romania, which gained Transylvania, which included a large Hungarian minority.

The Károlyi government preferred to resign rather than ratify this *diktat*. It was replaced the next day, March 21, 1919, by a new Bolshevik-inspired government, with Sándor Garbai as the official head of state, but with real power in the hands of the Commissioner for Foreign Affairs, Béla Kun. From the outset, he announced his intention to resist the demands of the Allies with arms and even to take back the abandoned territories, which initially won him broad popular support, including in military circles and even among the nationalist bourgeoisie.

Above: Having an aviation industry on its soil, the Hungarian government can take advantage of local production to perfect its air force - here, a UFAG C.I n°161.166 manufactured in Budapest (Coll. Bernhardt Totschinger)

Above: The Magyar Általános Gépgyár (MAG) factory in Budapest also produced fighters, such as this Berg D.I with the red star and Austro-Hungarian registration 92.126. (Paolo Varriale)

The weapons were therefore going to speak, but which ones could the Hungarian government count on? Like any belligerent in the Great War, Hungary had a number of experienced infantry troops. But, without a large national industrial base, the army was going to lack ammunition supplies, which came mainly from Austria and Germany. The Hungarian army was therefore forced to fight a short war... From the point of view of aviation, Hungary had an advantage over Soviet Russia in that it had factories producing modern war material. The Austro-Hungarian air force fought until the end of the war and was thus able to benefit from the latest and most advanced models designed by German industry and produced locally in three factories: the Flugzeuge der Ungarischen Lloyd-Flugzeugfabrik in Aszód, as well as the Flugzeuge des UngarischenFlugzeugfabrik A. G. (UFAG), and the Flugzeuge der Ungarischen Allgemeinen Maschinenfabrik (in Hungarian language *Magyar **Általános** Gépgyár* - MAG) in Budapest. Thus, Hansa-Brandenburg C.I or UFAG C.I two-seaters, and Aviatik-Berg D.I or Fokker D.VII fighters, all equipped with the powerful Daimler-Benz engine of 225 hp, were available.

As all the *Fliegerkompagnies (Flik)* of the Austro-Hungarian air force were stationed at the front outside of Hungary, the new Red Air Force (*Vörös Légjáró Csapatok*) could initially only rely on aircraft from the eight Austro-Hungarian flying schools (Flek) on its soil, i.e. about 150 aircraft, many of which were outdated and few of which were in flying condition. But the aircraft awaiting delivery in the three aviation factories were quickly added to the list. The Hungarian government ordered them to continue production in an order dated March 26, 1919. As of April 3, Lloyd was able to deliver 10 Aviatik-Berg D.I fighters that could be assigned to the front, in addition to those produced by UFAG and MAG, which also delivered some fifty excellent Fokker D.VIIs, whose production had just been launched when the war ended. These Fokker D.VIIs were generally preferred by Hungarian pilots over the Aviatik-Berg D.I.

This new air force, according to a report drawn up by the intelligence service of the French army in Hungary in June 1919, was directed by the Director General Samuelly with Lt-Colonel Petroczi as technical and military chief, assisted by Captain Steiner. It would have up to 8 operational squadrons with between 4 and 6 aircraft each: four stationed respectively in Györ (4th squadron, northwest of Budapest), Losoncz (today Lucenec, Slovakia – NE Budapest), Kecskemet (SE) and Kaposvar (1st squadron, SW), as well as four others in Bucharest itself, two of which were fighter squadrons including the 8th squadron, an elite unit where several veteran aces of the Austro-Hungarian air force such as Stefan Fejes (16 victories), Ferdinand Udvardy (9 v.), Johann Risztics (7 v.), and Alexander Kasza (6 v.). In addition to these ground units, Budapest also had a seaplane squadron and a flying school in Szombathaly, near the Austrian border.

Above: The best fighter available to the Hungarian communist government was the Fokker D.VII, of which an assembly line had been organized at the end of 1918 in the MAG factory.

Fokker D.VII of the 8th Vörös Repülöszazad in May 1919.

4.2 – The Romanian army settles on the Tisza (April 15 to May 3, 1919)

A first military adversary to the new regime, the French army, was already in the south of the country, in the Banat region, a region with a very heterogeneous population and a strong Hungarian minority, but which was assigned to the new kingdom of Serbs, Croats, and Slovenes (future Yugoslavia). The northernmost point occupied by the French troops was the Hungarian city of Szeged, but they did not go any further. Because the "French Army of Hungary" led by General de Lobit was composed of troops who only wanted to be demobilized and who were probably not ready to undertake a new military campaign – they would be replaced as far as possible by colonial troops. From the air point of view, an air park was set up further south at Novi-Sad (formerly Neusatz), which brought together escadrille 508 squadron and several Serbian squadrons. On March 14, 1919, the escadrille 508 received the order to move to Sezged, where it became operational on March 23 under the command of Lt Picot with only three Breguet 14s in flying condition, which immediately carried out reconnaissance flights towards the north. On March 25, 1919, Picot, with Lt Charmette as observer, flew over Bucharest and observed that there were no movements of the Hungarian army towards the French troops in the south, which, apart from a few incidents in the outposts, did not have to face the Communist troops.

Above: SPAD VII and Nieuport 27 fighters from escadrille 523 in Novi-Sad, 1919. (Coll. Boris Ciglic)

The main threat to the Hungarian government came from the east and the Romanian army. It was especially in front of communist Hungary that the bulk of the Romanian army would be concentrated, once again advised by the French, since General Berthelot returned to Romania to resume service with his staff officers. Unfortunately, this time he did not bring a cohort of French airmen with him, nor even any equipment, because the French *armée d'orient* had great logistical difficulties in convoying its own aviation and supplies from Salonika. Only one SPAD VII fighter was given as a token to the Romanian air force, which in March 1919 officially had 51 aircraft on the line, including some Austro-Hungarian army aircraft captured on the airfields of Transylvania and put back into service.

From the night of 15 to 16 April 1919, the Hungarian Red Army launched an assault on the new Romanian border after a short artillery preparation. The Romanian army, for a time pushed back, was able to organize a counter-offensive and push back, then pursue, its attackers, who were forced to retreat to establish a defensive line on the Tisza river, crossing Hungary from north to south, on 1 May. This was a leap of 200 kilometers to the west for the Romanian army, which was now only 100 kilometers from Budapest... The revolt began to rumble against Bela Kun, who had initially succeeded in attracting a great deal of sympathy among the population and in the army for contesting the borders established by the treaties. This defeat and the red terror he instituted made him lose it and he knew that his days were numbered if he did not obtain a quick military victory.

With its army installed on the eastern bank of the Tisza River, the Romanian air force only had three squadrons (N.7 fighter on Nieuports, S.2 and S.12 observation on Sopwiths) which totaled a meager handful of airplanes in flying condition, if one bases oneself on a French report dating back to the beginning of May and mentioning 2 operational airplanes on the Debrecen airfield, none on the Oradea airfield, and only one at Békéscsaba! The staff of the French *armée d'orient*, which had become the Danube Army, quickly delivered a first batch of 5 Breguet 14 A2s to the Romanian air force, which could resume its surveillance missions. A few skirmishes took place on the frontline when on May 19, 1919, a Romanian crew composed of Lt. pilot Ioan Sava Campineanu and Lt. observer Ioan Vlad, claimed a Hungarian two-seater near the village of Zagyvarékas. On June 12, 1919, it was the turn of a Romanian plane, a captured UFAG C.I piloted by S/Lt Stan Bucur with Lt Mihail Hurmuzescu as observer, which was shot down over the village of Sajoszöged. It was shot down by two Hungarian Fokker D.VIIs, the ace László Újvári (8 victories in 14–18) and his crewman Géza Keisz. A Hungarian two-seater wishing to flee the communist regime deserted the same day and took refuge in the Romanian lines.

During the night of April 15th to 16th 1919, the Hungarian Red Army launched an attack on the Romanian positions after a short artillery preparation. The attack quickly turned sour and led the Romanian army to organize a vigorous counter-offensive, which led it to make a leap of nearly 200 kilometers to the west, stopping on May 3, 1919 at the Tisza River, a tributary of the Danube that crossed the country from north to south. The capital, Budapest, was only a hundred kilometers away... Bela Kun's government lost a great deal of sympathy after this first military defeat, especially since the first measures of his policy aimed at nationalizing industries with more than twenty employees and collectivizing land without redistribution quickly alienated a good part of its potential supporters, while a red terror began to fall on the country.

Above: The new Czechoslovakian air force could organize a single squadron with former austro-hungarian planes, mainly Hansa-Brandenburg C two-seaters, seen here in Nitra in 1919 and used against the Hungarian Bolsheviks. (Coll. Berhardt Totschinger)

4.3 – Attack on Czechoslovakia (May 20 to June 30, 1919)

Bela Kun, after negotiating a fragile cease-fire with the Romanian troops, sent his army on the attack on May 20, 1919, heading north against Slovakia and Ruthenia, which had been attached to the new Czechoslovak state. For the Republic of Councils, the conquest of Slovakia had a political objective because the south of the province was mainly populated by Hungarians. The conquest of Ruthenia, populated by Ukrainians, was above all an objective of political survival: it would allow to make the junction with the Bolshevik troops of Soviet Russia.

Faced with the Hungarian Red Army, the troops of the young Czechoslovak nation were ill-prepared and would soon have to give up ground. From the air force point of view, its airmen, all veterans of the Austro-Hungarian air force, would have to fight against their former colleagues serving in the Hungarian air force! They could only count on a stock of Hansa-Brandenburg C.I. two-seaters. The contribution of the French military mission present since April 9, 1919 on the field of Kbely near Prague would be determining: from its instructors and its aircraft delivered in boxes (Breguet 14, Salmson 2A2, SPAD VII and XIII) would be formed the BR 590 squadron which would be led by a veteran of the anti-Bolshevik war, Captaine Georges Lachmann returned from his adventures in Arkhangelsk of the previous year, and still as motivated to take up the fight at the controls of his personal SPAD XIII that he decorated with his famous question mark. He had under his command a number of particularly motivated French pilots, each of whom flew a Breguet 14, and he took five of them to the Vajorny airfield near Bratislava on June 6, 1919. They included S/Lt Jean Dupré, Adjudant Honoré Soubira, as well as Sergeants Georges Pariset, François Forgues, and Joseph Pouliquen. Pouliquen, who was to become commander of the Normandie-Niemen fighter group in 1942, gave an account of his experience:

"After the war [of 14-18], we received circulars in the squadrons. They didn't really know what to do with us. There were military missions that were being created everywhere. I volunteered and I was taken to Czechoslovakia (...) I fought against Bela Kun, against the Bolsheviks, which gave me my TOE war cross of which I am very proud because few reserve officers have it. (...) I wanted to continue to fly, we were told that there would be a fight against the Russians.

We were housed at the Carlton hotel in Bratislava. Our pay was higher than that of the Czechoslovakian air minister. And we had the prestige of the French, despite the fact that we were on foreign territory. Because we passed Hungarian officers in the street who looked at us as if we were winners and did not like us very much. When we entered the theaters to see the operettas, they were close to fight against us, nudging us, especially if we were with a pretty girl... I remember that I had brought a pair of red infantry breeches from France, without the black officer's stripe. I never wore them, but one day when my uniform was at the dry cleaner's I put on these red breeches to go to the field. To get to the park where the cars were, we used to stop in a bakery to eat cakes. From that day on, everyone took out theirs, it was the Empire [of Napoleon], it was the French Army! When I left, my comrades asked me to leave them these red breeches, because they took turns putting them on when they went to see a girl..."

As soon as he arrived with his men, Lachmann left the command of BR 590 to Lt Roussin because he was appointed head of aviation on the Slovak front, which also included the Czech squadron flying Hansa-Brandenburg C.I

Above: Phönix C.I were also used by the Czechs. (Coll Berhardt Totschinger)

Above: Breguet 14 A2 of French escadrille 590. (Coll. Michal Pavlec)

Above: Sergent Joseph Pouliquen seen here at the commands of his Breguet 14 A 2 in Bratislava in 1919.

Above: French pilots of escadrille 590 decorated during a ceremony after the campaign against Hungary – Sergent Joseph Pouliquen under the cross.

Above: French planes given to the young Czechoslovakian air force – SPAD fighters and a Salmson 2A2 at the center of the photo, even if this two seater was not put into service against the Bolsheviks Hungarians.

from the stocks of Austro-Hungarian aviation. The French pilots, who had no licensed observers or gunners, routinely flew with their mechanics sitting in the turret, or with Czech officers.

However, faced with the push of the Hungarian Red Army, the small Franco-Czech air force could not perform miracles, but its aircraft proved to be quite useful in observing enemy movements. Thus, on June 20, 1919, the air force noticed that the city of Bratislava was not threatened by the enemy army, allowing Czechoslovak troops to redeploy accordingly. The Breguets remained effective in bombing missions on railway stations and tracks to disrupt enemy supplies, even if the crews had to improvise to compensate for the lack of equipment, as Joseph Pouliquen testified: "*There were no bombs under the wings because we had no bomb-dropping system. We threw Austrian bombs by hand. But we could aim right because we were passing 100 meters above a train or above a marshalling yard, and the observer or the mechanic – mine was called Coutème – would take the wing bomb with a handle and drop it. We got caught in the wings several times with machine gun bullets, and I came back once or twice like a sieve. The trick was not to get shot in the butt or the skull!*"

The Breguet 14s were to intervene on the towns of Komarno on the Danube, as well as Nové Zamky and Levice further north, providing significant fire support to the Czech troops fighting there and demoralizing the enemy troops. The station of Nové Zamky was bombarded several times on 7 June and the Bolsheviks withdrew. To increase the effectiveness of the strikes, three auxiliary fields were used at Horny Bar, Nemes Ocsa, and Nitra.

The Hungarian Red Air Force was very active, especially its 4th squadron based at Gvör, not far from Bratislava, which had three to four Aviatik-Berg D.I and several Fokker D.VII. The Hungarian top ace Stefan Fejes, a veteran of the Austro-Hungarian air force where he had won 16 victories, broke the propeller of his Fokker D.VII at the end of May 1919, during a strafing mission on the ground, and was captured. On June 14, 1919, the only air combat of the campaign took place when a Breguet was attacked by an Aviatik-Berg D I (n°348.46) piloted by Joszef Matyasffy who claimed victory, although the French aircraft piloted by Sgt Forgues returned to its base – the Czech machine gunner Robert Ellner was awarded the Czech War Cross for resisting the attack. On 15 June, most of the French aircraft took part in the battle around the town of Levice with the Czech Hansa-Brandenburg C.I, without being able to provide decisive support. On June 20, 1919, they had to defend the town of Nové Zamky, an international incident occurring when a mineral oil refinery belonging to Standart Oil was bombed by mistake – probably by the Pouliquen / Ellner (Czech) crew.

The action of the Breguet 14s annoyed the enemy air force to such an extent that it organized a daring raid on June 21 against the advanced airfield of Nemes Ocsa. 6 Fokker D.VII and Aviatik-Berg D.I fighters escorted a single Hansa-Brandenburg C.I carrying two pilots as passengers, who were supposed to seize two Breguet 14s on the ground to fly them back to Hungary... They only discovered an empty field and returned empty-handed, taking only the cloth landing "T" as booty.

Above & Below: These photos of the wreck of SPAD VII seen in May 9, 1920, well after the war against Hungary, are very interesting because showing plane of the French ace Georges Lachmann, decorated with his personal marking, the question mark.

Breguet 14 A2 n°9122 of sergent Joseph Pouliquen, escadrille 590, Bratislava 1919.

SPAD VII (serial unknown) of Capitaine Georges Lachmann, escadrille 590, Bratislava 1919.

The conflict ended with a cease-fire by Bela Kun on June 24, 1919, when the French government threatened a joint attack by Romanian, French, and Serbian troops in the south and east of the country. The Hungarian troops retreated under the surveillance of the Breguets of escadrille 590 towards the negotiated demarcation line. During this short campaign, the Br 590 squadron, with 7 Breguets and 2 SPADs, carried out between 13 and 30 June 1919 a total of 15 reconnaissance flights, 28 bombing raids (136 bombs launched) in 38 flights, supported by the Czech squadron over Hansa-Brandenburg which carried out 11 flights during which 42 bombs were launched.

4.4 – The final assault of the Romanian army (July–August 1919)

Bela Kun did not succeed in making a junction with Soviet Russia and prepared his last stand by planning an attack against the Romanian army on the Tisza.

Other air incidents took place in the southern sector held by the French troops of the Hungarian army, bordering the city of Szeged, where France installed a counter-revolutionary Hungarian government on May 30, 1919, presided over by Gyula Peidl, but whose strongman was to be the Minister of War, Admiral Miklós Horthy. Since May 11, 1919, the Franco-Serbian fighter escadrille 523 (equipped with SPADs and Breguets 14) had come to reinforce the 508 at Szeged, and every day the two units sent their Breguets, as well as an older Dorand AR, to carry out reconnaissance on the enemy lines, which took place without incident. The atmosphere in the enemy air force seemed to be very unhealthy, because on June 14, 1919, a new aircraft deserted and landed at 8 a.m. in Szeged, where its pilot took himself prisoner, declaring that he wanted to flee from the Reds... The Communist air force dropped a few leaflets over Szeged, inciting the population to rise up, and on June 15, General Franchet d'Esperey telegraphed strict orders to the French SPADs to shoot down any enemy aircraft that risked such a mission. This was not to be, as the conflict was soon to end.

On July 20, 1919, the Hungarian communist regime played its go-ahead and launched its troops to attack the Tisza River, which they managed to cross at four points, putting the Romanian army in difficulty. But the latter

Above: Three French pilots pose in front of a SPAD VII delivered to the Romanian aviation, decorated with the name Marasesti – the famous Romanian battle of year 1917.

End of the line for this UFAG C.I of the Hungarian Bolshevik Air Force.

quickly counter-attacked on 25 July, supported by its air force, which had 17 aircraft in flying condition thanks to French deliveries. Within a week, the Romanian army regained its starting positions and crossed the Tisza River, resulting in the last air battle of the war on July 31, 1919, when a formation of Hungarian aircraft (Hansa-Brandenburg C.I.) attempting to bomb the Romanian army's pontoons was intercepted by the Nieuports of the N.7 squadron one of which, piloted by Lt. Iosif Răcăşanu, claims to have shot down one of the attackers who fell into the enemy lines at Tiszalök.

Barely a week later, on August 6, 1919, the Romanian army entered Budapest and precipitated the fall of the Hungarian Communist regime, whose leader Bela Kun managed to flee the country and went on to serve as an executive of the Russian Communist regime, "distinguishing" himself by overseeing the execution of the White Army cadres captured in the Crimea at the end of 1920 – he himself would be executed during a Stalinist purge in 1938. The provisional Hungarian government of Szeged, of monarchist inspiration, took power in Hungary and, more concerned with the anti-communist struggle than with the reconquest of the former territories, agreed to sign a peace treaty with Romania, a treaty that provided for the transfer to the victor of its air force, which included about 150 aircraft of the former Austro-Hungarian air force, 107 of which were more or less in flying condition. It was also at the same time that the French army of the Danube evacuated the country, leaving the Romanians all its stocks of Breguet 14 and SPAD aircraft.

On August 9, 1919, the only loss of French aviation during the campaign occurred when a Breguet 14 of the 523 squadron, piloted by Lt Robert Gaborit and carrying as observer S/Lt Pierre Paul Marguerite, crashed in Szeged during a test flight, killing its crew. The Bolshevik threat being over, the French army in Hungary was dissolved.

Above: Salmson 2A2 n°43xx of escadrille SAL 582 in Poland, 1919. Like all Salmsons of the serial 4000, this plane was built in the Latécoère factory of Toulouse.

4.5 – The French military mission in Poland

As communist Hungary collapsed, another European nation, Poland, fought for its independence. One of the consequences of the First World War was the rebirth of Poland, a nation that had been torn apart at the end of the 18th century and divided between its three powerful neighbors, Russia (center and east - Grand Duchy of Warsaw), Austria-Hungary (south, Galicia), and Germany (west, the so-called Greater Poland region). During the war, Polish soldiers served under several flags; Russia and Austria-Hungary even authorized the creation of Polish armies within their forces, especially the Polish Legion led by Jozef Pilsudski who fought in the Austro-Hungarian army under the promise of the German and Austrian governments to create an independent Polish state after the war. In the conquered areas of the Russian Empire, Germany and Austria-Hungary effectively created a "Kingdom of Poland" in 1916, a semi-independent state in which they reserved areas of influence, but which in fact became a satellite of Germany. The merit of this puppet state was that it laid the foundations for a Polish administration. In France, Jozef Haller, a former senior officer in the Austro-Hungarian army who had gone to Russia, formed a legion of Polish volunteers recruited from among the emigrants, known as the "blue army", which went to fight on the Western Front.

At the time of the armistice of November 11, 1918, which saw the collapse of the German and Austro-Hungarian empires, the Kingdom of Poland, a satellite of Germany, served as a base for the Republic of Poland, which was proclaimed on its soil, with General Jozef Pilsudski becoming its head of state and military leader. With no clear borders defined in areas where several nationalities lived, the young Polish nation immediately had to fight and set about creating armed forces, including a military air force (*Lotnictwo Wojskowe*).

At that time, Poland had a number of experienced pilots who had served in the Russian air force, where Polish squadrons were created at the end of the war. Among the 13 fighter aces of the Russian air force that survived the war, there was even a Polish pilot, Poruchik (Lieutenant) Donat Aduiovich Makijonek (8 victories), who immediately joined the new air force. There were also a number of Polish pilots who had served in the Austro-Hungarian air force and, to a lesser extent, in the German one. Finally, there were pilots in France, most of whom were still in training, who were to serve as the air component of General Haller's Blue Army.

The equipment was acquired by seizing stocks of Austro-Hungarian or German aircraft present on the territory of the new Polish Republic. The air base of Mokotow, near Warsaw, was seized on November 15, 1918, after its evacuation by German troops, allowing the recovery of nearly one hundred German aircraft, mainly two-seaters. Those that were in flying condition were immediately put into service and decorated with a white and red shield as national marks. At the same time, the Poles of Galicia took over the Rakowice airfield, near Krakow, where there was an Austro-Hungarian air school, allowing them to recover several Brandenburg C.I, Oeffag, and Lloyd two-seaters, which were put into service with a white square insignia marked with a red "Z". In total, according to a count dated January 26, 1919, 237 German and Austrian aircraft were seized, but only 45 were in flying condition. A new stock was soon added to this total following the insurrection launched in December 1918 in Wielkopolska (Greater Poland), an officially German region located to the west of the Kingdom of Poland. In this new territory, seized by Polish troops, there were nearly 400 German airplanes in the airship hangar of Winiary, near Poznan, which were generally out of service but which would

Above: Polish pilot in front of his SPAD XIII just delivered by the French. On the foreground, a Breguet 14.

provide material for mechanics to restore aircraft to flying condition after the purchase of spare parts in Germany.

This new *Lotnictwo Wojskowe*, as soon as it was formed, was immediately involved in the conflict that broke out with the People's Republic of Western Ukraine (ZNUR), which had made the city of Lwow its capital (today Lviv, Ukraine). The Polish population of the city, which had a numerical majority, did not accept this and went into insurrection on November 1, 1918. The local Polish airmen managed to seize the Lawandoks airfield, still held by Austro-Hungarian troops, where they were able to seize a few Brandenburg C.I. two-seaters on which they painted red and white stripes on the wings and tailplane. These aircraft were the only link between the Polish insurgents and the outside world, as Ukrainian troops soon surrounded them in part of the city.

A generalized conflict then began, which saw the arrival of Polish troops from Galicia, who managed to deliver the insurgents from Lwow on November 21, 1918 by opening a narrow corridor that was still threatened by Ukrainian troops. The latter had a small air force to support their soldiers, made up of a few Russian (Nieuport) and German aircraft recovered from local airfields. For its part, the *Lotnictwo Wojskowe* was structured and adopted at the end of December 1918 a unique insignia on its aircraft, replacing the three different marks adopted locally in Warsaw, Krakow, and Lwow: the famous red and white checkerboard, which was the personal insignia of Lt. Stefan Stec, a veteran fighter pilot of the Austro-Hungarian air force who now served in the Polish air force, where he was entrusted with one of the rare fighter aircraft in the inventory, a high-winged Fokker E. V high wing monoplane. In January 1919, the Lotnictwo Wojskowe had seven squadrons (eskadra) of six two-seater aircraft of different types, divided into three groups located in Warsaw and Lublin (1st, 2nd, 3rd, and 4th Eskadra), Cracow (5th and 6th) and Lwow (7th). In these three cities, military workshops were busy refurbishing as many wrecks as possible from the captured stocks. In February, four other squadrons were created in Wielkopolska (Greater Poland), in charge of monitoring the border with Germany, where several incidents took place against Freikorps.

In March 1919 Polish troops, led by General Edward Rydz-Śmigły, went on the offensive and broke through the Ukrainian front. The Polish air force supported the attack to the best of its weak means with the 3rd Eskadra commanded by ace Donat Makijonek in Lublin and then Kowel and the 5th and 7th Eskadra in Lwow. During these fights, on April 29, 1919, Lieutenant Stefan Stec, of the 7th Eskadra, won the first aerial victory of the Lotnictwo Wojskowe on board his Fokker E.V (n°187/18): he intercepted a Nieuport (n°3598) of the 1st Ukrainian squadron escorting a two-seater, concentrating his fire on the fighter which fled in the fog. The Polish infantry confirmed the plane as destroyed but this one in fact managed to return to its field, damaged. Stec was unaware that he was in fact fighting a former comrade of the Austro-Hungarian air force: the ace fighter Franz Rudorfer (11 victories), who came to fight as a mercenary in the Ukrainian air force! Stec won another victory against a captive balloon on May 10, 1919, his second success in the

Above: Breguet 14 A2 with French cockades seen in Poland on July 1919.

Above: Change of owners : this mechanic paints the polish chessboard on the wing of a Breguet 14 in 1919.

Polish air force in addition to the three others in the Austro-Hungarian air force, making him an ace.

During this period, the Polish army received a very important reinforcement in the form of the Haller army, also known as the blue army, which arrived by rail from France and whose soldiers were engaged on the Ukrainian front. In addition to being a well-trained and seasoned unit, the Haller army also had the advantage of arriving with an air component consisting of 7 squadrons whose pilots were mostly French: SPA 162 fighter (15 SPAD VII), BR 39 and BR 59 reconnaissance (10 Breguet 14 A2 each), BR 66 bombing (15 Breguet 14 B2) as well as SAL 580, 581 and 582 for artillery spotting (10 Salmson 2A2 each), a total of 101 aircraft including those brought in as reserves.

The French pilots, who were not under the Polish command but under that of the French mission in Poland, were mainly tasked with training their Polish colleagues in the operational use of their aircraft; nevertheless, BR 59 and SAL 581 were to be employed on the Ukrainian front at the Volodymyr-Volynskyï airfield from May 1919 onwards to

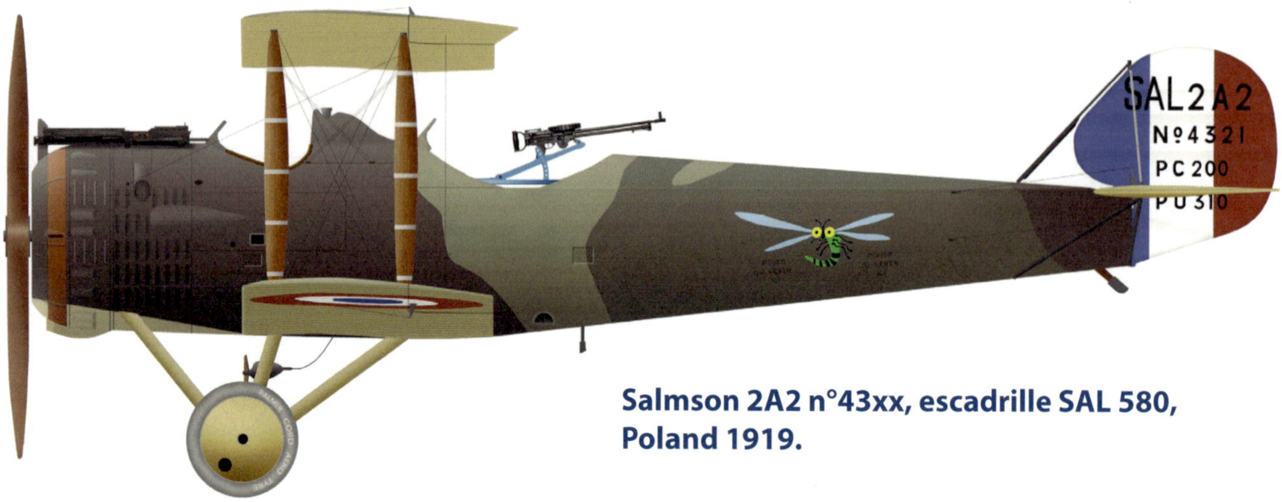

Salmson 2A2 n°43xx, escadrille SAL 580, Poland 1919.

support the progression of the Haller army towards Lutsk, while a detachment of the BR 39 based in Lublin was also to take part in the conflict. The Ukrainians counter-attacked successfully from June 8 and pushed back the Poles over a distance of 120 km, but the latter, under the command of General Pilsudski, established the situation from June 27, conquering all of eastern Galicia, which was de facto annexed after the cease-fire signed on July 17, 1919.

While the Polish Army was fighting against the People's Republic of Western Ukraine in the south-east of its borders, another front openned against the Red Army which, wishing to take back the territories evacuated by the Germans, advanced towards the west and tried to invade the three Baltic states that are Estonia, Latvia, and Lithuania, and in so doing approached the still ill-defined limits of the new Poland, seizing in April 1919 the cities of Vilnius and Grodno, which were located in regions with a majority of Polish inhabitants. At the same time, General Pilsudski wished to extend the limits of the new Republic of Poland and ordered his troops to advance towards the north-western territories, towards Lithuania, where battles were to take place against the Red Army, to whom the city of Vilnius was taken back on April 19, 1919, as well as towards Belarus, where the city of Minsk was conquered on August 8, 1919.

The *Lotnictwo Wojskowe* took part in these battles and the French squadron SAL 582 was engaged in the front line by setting up on the Vilnius airfield at the end of May 1919. The unit conducted several war missions from this airfield, in particular long-range reconnaissance missions on the Bolshevik lines; during one of these missions, carried out over the city of Smolensk on June 12, 1919, a Salmson 2A2 flown by a Franco-Polish crew was shot down by Bolshevik flak and had to land on enemy territory. The Polish pilot, Lt Wiktor Wolski, was interrogated by the Bolshevik police and summarily shot, as was probably the case for the French observer, Second Lieutenant Raoul Marcel Fourrey (23 years old), who died in Smolensk on August 19, 1919, according to French archives, and was probably the only French airman killed in action during the fight against the Red Army. Some inconsequential clashes took place between Polish two-seaters and Nieuport fighters of the RKVVF (Aviation forces of the red army), several of which deliberately deserted in the direction of the Polish lines at that time.

The fighting stopped towards the end of the summer of 1919 because of the offensive of Denikin's White Army towards Moscow and because of General Pilsudski's desire not to favor the latter, as Denikin had refused to recognize Poland's independence and wanted to re-establish Imperial Russia within all its borders. However, Pilsudski knew that sooner or later his country would have to face Russian imperialism again, no matter which side would take over the Kremlin, and he was effectively led some six months later to resume hostilities against the Bolsheviks by making peace with his former enemies, the People's Republic of Ukraine led by Symon Petliura. At the beginning of 1920, the latter saw the territory he controlled reduced to a trickle by the advance of the Red Army and proposed to Pilsudski a military alliance to push back the Red Army towards the east. The Polish general, who hoped to form a federation with Ukraine and the Baltic states to resist Russian appetites, accepted and hostilities resumed in April 1920 with the aim of taking the city of Kiev.

The *Lotnictwo Wojskowe* took advantage of the winter of 1919–1920 to expand its ranks in terms of equipment and pilots. On September 25, 1919, Colonel Gustaw Macewicz was appointed head of the air force and he merged the four squadrons of the Wielkopolska and those of the Haller mission into the *Lotnictwo Wojskowe*. The former became the 12th, 13th, 14th, and 15th Eskadra, while some of those of the Haller mission became part of the Polish air force as a formed unit, being renumbered as the 16th (ex-BR 39), 17th (BR 59), 18th (SAL 580), and 19th (SPA 162) Eskadra. The other French units were disbanded with their aircraft assigned as reinforcements in various units. In the spring of 1920, there were 20 Eskadras on the line, four of which specialized in hunting (7th, 13th, 15th, and 19th Eskadra Myśliwska) and one in heavy bombing (21st Eskadra Niszczycielska). There was no squadron with the number 20.

Polish aviation had to face many fights during the year 1920. The Polish army first took advantage of the Russian civil war to advance eastwards and even took the city of Kiev in June 1920, before retreating in the face of the Red Army, which, freed from the threat of the White Armies, invaded Poland and arrived at the gates of Warsaw in August 1920, before being chased away by a Polish counter-offensive which restored the front line to the starting positions of 1920, the conflict ending in the autumn. This war did not concern the French pilots who all returned to France and is not the subject of this book... But it is worth mentioning that a contingent remained in Warsaw to organize the flying school that opened there, which proved to be a decisive help in training a total of 43 Polish pilots until the autumn of 1920, each one with nearly 120 hours of flight time to his credit – a very advanced training compared to the other school operating in Krakow according to Austro-Hungarian methods and which trained 83 pilots during the same period but with only about 30 hours of flight time per student.

Chapter 5: Colonial Fronts

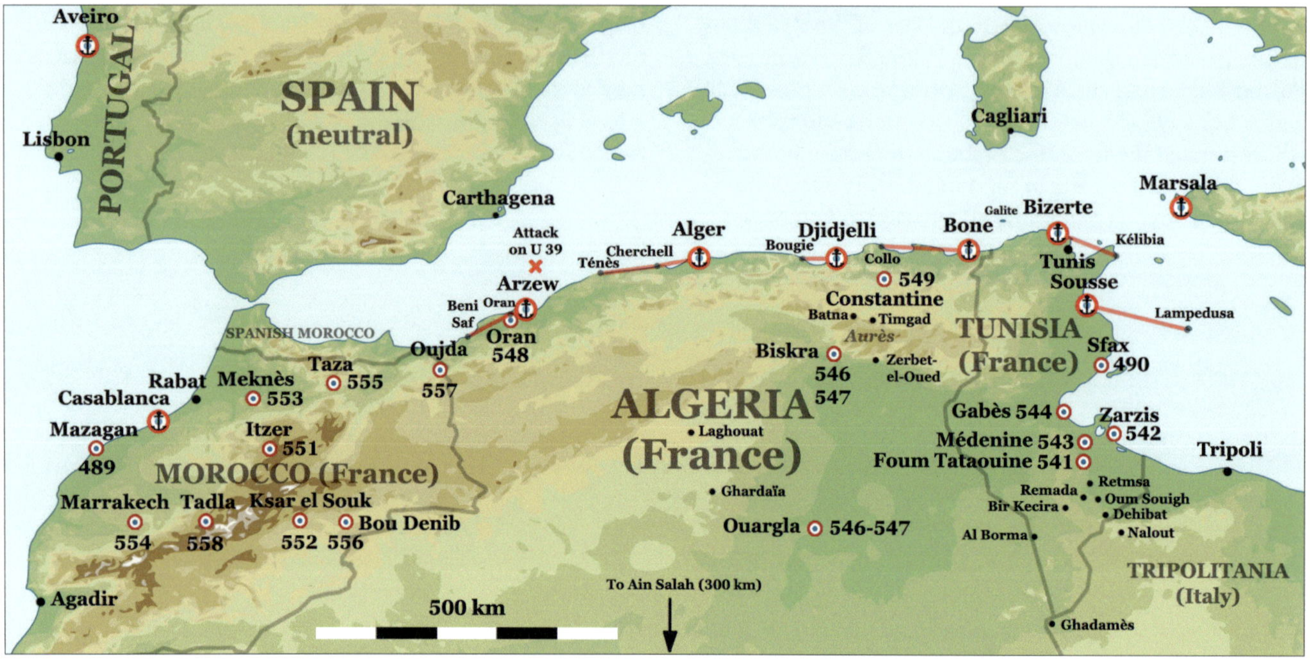

5.1 – Military Aviation Units in North Africa before WW1

At the beginning of the 20th century, France had a vast colonial empire in Africa, particularly in North Africa. Algeria was the oldest of the French colonies, whose conquest began in 1830; it became a settlement colony with a large number of colonists from France, who made up about 15% of the population in 1914. The entire coastal strip of the country was pacified, but the vast area of the Sahara remained largely untamed and administered directly by the army. Tunisia was conquered in 1881 at the expense of the Ottoman Empire and was officially a protectorate, although in fact the country was managed as a colony by the French authorities; French settlers represented less than 5% of the population at the time. Control of Morocco was very recent, with the establishment of a protectorate imposed on the sultan in 1912 and a military conquest led by General Lyautey, which met with strong resistance to the point that many areas remained rebellious to French authority.

When the war broke out, almost all of North Africa, with the sole exception of Spanish Morocco, was controlled by France and constituted an important military resource, allowing the recruitment of colonial troops. The German army's General Staff was aware of this and tried everything to prevent France from taking advantage of it, first by trying to disrupt maritime navigation between Algiers and Marseille with the help of its ships of the line *Groeben* and *Breslau*, which were supposed to prevent the convoying of North African regiments to the front. This plan turned out to be a failure, as the two ships, after having fired a few shots at the Algerian coast, were forced to take refuge in Turkey without being able to disrupt the flow of troops to Marseille. Germany had one other means left: to encourage the indigenous insurrections, which were to take place with varying degrees of success. To counter them, France was to use a new military means that allowed it to maintain its authority at little cost: aviation.

Aviation in North Africa was born at the same time as aviation in metropolitan France, because the French population of Algeria provided its number of pioneers. Louis Mouillard, an ornithologist and farmer in the vicinity of Algiers, made a light glider comparable to that of Otto Lilienthal in 1865 and made a short flight of 42 meters in 15 seconds from an embankment, where he damaged his shoulder on arrival. He realized the first recorded human flight in Africa... There were several balloon ascents at the end of the 19th century and on November 18, 1909 the pioneer René Métrot realized a one kilometer flight on his biplane Voisin on the Caroubier racetrack, next to Algiers (in Hussein-Dey), which constitutes the first recorded flight of a motorized plane in Africa. Other pilots followed and the Algerian public was enthusiastic about the aviation week organized from May 12 to 19, 1912 at the maneuvering field of Algiers, where they could admire a tethered balloon and several aircraft.

In 1909, the government and the army saw the value of aviation for the exploration of the Sahara and the establishment of a postal service in remote areas. On the decision of General Roques, Minister of War, Lieutenant Maxime de Lafargue, an officer of the 3rd Spahis regiment stationed in Algeria, was sent to France to take flying lessons and obtained his civilian brevet on March 3, 1911 on a Henri Farman. The French parliament voted credits for the creation

Above: Propaganda postcard for military aviation in Africa.

of a squadron in Biskra, a city on the edge of the Sahara and connected to the coast by a railway line. The small formation was created at the beginning of 1912, when Maxime de Lafargue made his first flight on a Farman type 1911 on February 17, in front of the entire local population, which was strongly impressed. The pilot underlines "*the considerable moral effect produced by these first flights on the Arabs and, in particular, on the Nomads of the south, who prostrate themselves at the passage of the aircraft.* He made other flights around Biskra with his handful of pilots and organized raids to other oases of the Sahara, such as Touggourt and Ouargla, not without difficulty because he had to face the desert winds. On February 26, 1913, four Henri Farman HF 20 of the squadron, piloted by Lieutenants Ernest Reimbert, Etienne Cheutin, Joseph Jolain and *maréchal des logis* Pierre Hurard, taking on board their mechanics as passengers (including Corporal Emile Dewoitine, the future famous aircraft manufacturer, at the time Reimbert's mechanic) took off from Biskra and flew to Tunis after several stops in Tozeur, Gabes, and Sfax. In July 1913, an aviation center opened in Manouba, north of Tunis, commanded by Lieutenant Reimbert, who proposed the organization of a new prestige raid from Tunisia to Morocco. General Lyautey, head of the army in Morocco, refused, arguing that he did not want the airmen to run the risk of falling into a rebel-held zone following a mechanical breakdown! Reimbert postponed his ambition to a five-aircraft raid that led him from Tunis to Oujda, in the center of the Sahara, which his pilots reached on May 11 after five days of flight in stages during which nearly 3,000 kilometers were flown in 43 hours of group flight, being preceded by a car convoy that ensured their refueling and breakdown services.

Lyautey, for his part, was not uninterested in aviation... He had two squadrons at his disposal which were used for the conquest of Morocco. The French Army progressed along two axes converging towards the center of the country: one from the city of Casablanca on the west coast, and the other in the east from the Algerian border. An air center was planned to accompany each of these French armies. The first one left Marseille for Casablanca on February 12, 1912, equipped with four 1911 Blériots under the direction of Lieutenant Pierre Clavenad, an infantry officer who held a military pilot's license number 9. The second one moved to Oujda in May 1912 under the command of Lieutenant Henri Jannerod, but had to wait until January 1913 to receive its 4 Deperdussin. The stewardship was so poorly organized that the airmen were practically left to their own devices. Lieutenant Clavenad, even before his airmen were able to

Above: Blériot monoplane on the beach of Casablanca, during spring 1912.

Above: The Blériot monoplanes of the Casablanca escadrille began operational flights in late August 1912 and had a complete operational use in October, making aerial reconnaissance for French infantry moving eastward during the conquest of Morocco.

carry out their first missions, decided, with the agreement of General Lyautey, to return to France to investigate the origins of this organizational disaster. He then alerted public opinion by suggesting an article published in the newspaper *Le Matin* on July 15, 1912, which simply informed the public that a military air force was fighting in Morocco at the cost of many difficulties. He was much more explicit five days later when he personally signed an article entitled "Le cri du soldat" (The cry of the soldier) in which he denounced, without taking the gloves off, all the negligence of the military administration which condemned his unit to inaction. He was transferred and punished as he had expected, but his daring move helped to unblock the supply situation and the first operational flight was accomplished on August 18, 1912 with a reconnaissance mission towards rebel groups in the Séfrou region. The operational activity of the Casablanca center did not really begin until October 1912, when the Blériots regularly accompanied the columns of French troops as they advanced and warned them by weighted messages of the presence of rebel troops. At the beginning of 1913, while the Oujda center was also making its first military flights on its Deperdussins, Lieutenant Van Der Vaero, who commanded the Casablanca center, sent a report to France in which he asked for bombs, a wireless device, and cameras for his aircraft. In return, he received an insulting letter written by an assistant to General Hirschauer, then head of the military air force, telling him that the African sun had beaten down on his head and reminding him that "the fate of Alsace-Lorraine was not at stake on the Moulouya river". It is true that the war was approaching...

Above: Autochrome showing a Deperdussin monoplane of the Oudja escadrille during the conquest of Morocco, 1913. (Jules Gervais-Courtellemont)

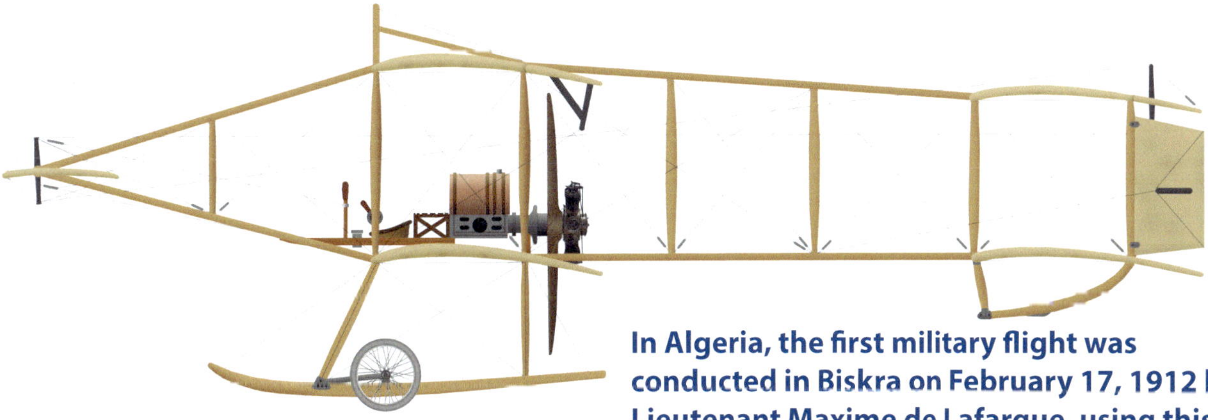

In Algeria, the first military flight was conducted in Biskra on February 17, 1912 by Lieutenant Maxime de Lafargue, using this Farman type 1911.

and all the new aviation equipment was being used in metropolitan France. On May 1, 1913, the Minister of War, Eugène Etienne, sent Captain de Chaunac to inspect the Moroccan air force. He returned to France on September 9 and his report deplored the poor use of intelligence transmitted by the air force, which should nevertheless be maintained and better organized with the use of the wireless.

When the First World War broke out at the beginning of August 1914, the French army had largely experimented with aviation in the fight against indigenous rebels, which it used to recognize positions but also for bombing. It also had a small core of pilots experienced in desert flying... All these pilots and equipment were sent to France to fight against the German army, so that the North African forces had practically no more aerial means in 1914. But the development of colonial rebellions would change the attitude of the French General Staff in 1916.

Left: The first airplane to reach Biskra by train in 1911, this Maurice Farman type 1911 with 50hp engine. The pilot in the picture is Lt Jolain, on the camel is Lt Cheutin. (Collection Pierre Jarrige).

Right: Lt de Lafargue, commanding officer of the Saharan escadrille of Biskra, flies above the oasis in 1912. (Collection Pierre Jarrige)

Left: Corporal Emile Dewoitine, future famous French aviation constructor, was in 1913 an ingenious mechanic serving in Biskra. He conceived this car equipped by a propeller and powered by the engine of a broken Farman… (Collection Pierre Jarrige)

Right: A Henri Farman HF 20 taking off from Biskra in 1913. (Collection Pierre Jarrige)

Left: Henri Farmans HF 20 from Biskra, during their propaganda raid across North Africa, seen here in Tozeur (Tunisia) on February 27, 1913. (Collection Pierre Jarrige)

Right: The Henri Farmans from Biskra are now in Colomb Béchar (Algeria, near Moroccan border). (Collection Pierre Jarrige).

Above: Lieutenant Etienne Cheutin was one of the first military pilots of the French air force. He conducted a raid from Biskra to Tunis with 3 other pilots on February 1913 at the commands of a Henri Farman HF 20. In 1917, with the rank of commandant (Major), he was named commanding officer of the aviation in Morocco. He ended his career in 1934 with the rank of General.

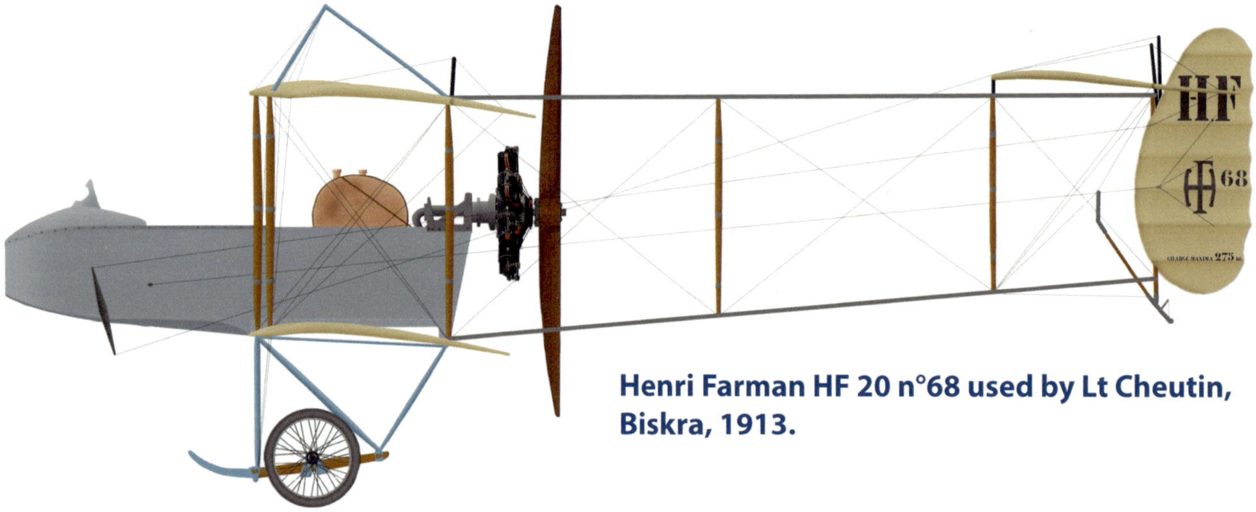

Henri Farman HF 20 n°68 used by Lt Cheutin, Biskra, 1913.

5.2 – Aviation in Morocco 1914–1918

Since the establishment of the protectorate in 1912, Morocco has always been ruled by General Hubert Lyautey, with the title of resident-general. When the war broke out, not only his air force, but also a large part of his military resources were withdrawn, to be sent to the French front. His policy of civil concord and economic development allowed him to hold the country during the first years of the war, but the situation did not last as unrest broke out, as he explained in a letter dated March 2, 1916, which he addressed to the Minister of War and which perfectly summarizes the situation:

"The latest military operations in Morocco have made it possible to note among our adversaries improvements in organization that should be of concern. Thus, among the objects that our troops seized in the camp of Abd el Malik on January

Above: Detachment of escadrille F.552 in Sidi Lamine, Morocco, February 1918.

Above: Pilots and observers of escadrille F.554 pose in front of one of their Farman F.40s, Marrakech, Morocco, circa 1918. Note the Fatima hand in a moon crescent drawn on the plane.

26 last, were the elements of a T.S.F. post. On the other hand, the djich [armed band, author's note] *of the Beni Bou Yahi and Métalsa, which cut the railroad tracks from Taza to Oujda, on February 18 last, made use of articulated saws of various models and dynamite or cheddite firecrackers. This modernization of the warfare methods used by the Arabs until now was probably carried out by the Germans, who got in touch with the leaders of the hostile Harkas; in addition to its material effects, it is likely to raise the morale of the Harkas and incite them to a redoubling of their aggressiveness, which it is important to prevent without delay.*

To this end, it is necessary to reinforce the material and moral superiority that we already possess by using new means of action, the most important of which is aviation. This is, in fact, the field in which we will be able to maintain this superiority for the longest time, because as long as the control of the seas belongs to us, it will be almost impossible for the Germans to bring to the hostile tribes the assistance of their aeronautical service. The intervention

Above: Farman F40 (n°4414) of escadrille F.558, Tadla, Morocco, 1918. Barely visible within the circle on the side of the fuselage is the squadron symbol, an owl perched on a branch. The skull and crossbones on the front of the fuselage is a personal insignia.

of the air force will have material and moral effects of the first importance. From the material point of view, the airmen will be excellent agents of liaison, exploration and execution. The liaison between the columns has become more necessary than ever since Morocco has sent most of its active units to the North-Eastern borders. Those remaining in Morocco have been organized in mobile groups intended to act simultaneously on the objectives to be reached; the concordance of their efforts requires a close liaison

Above: A less successful version of the aircraft, the Renault 8Gc 170hp-powered Voisin VII was sent to North Africa, seen as a secondary front. This plane was used by commandant Cheutin to cross the Atlas mountains from the airfield of Meknes to the city of Bou Denib in June 1917.

Above: Voisin VII from escadrille 302 in Taza, on December 1916.

Above: Escadrille F.556 owned a Nieuport section. One of these planes visited Colomb Béchar in 1918. (Collection Pierre Jarrige). (Coll Jack Herris)

between them which can only be fully achieved by aircraft. It is also only on the aviation and its photographic processes that the command can count to have a precise knowledge of the ground and of the enemy groups, especially since the armament of the latter has been perfected and that the operations continue in the mountainous regions of Morocco. Finally, the bombardment of these groups or the douars that shelter them by aircraft will begin their disintegration and weaken their subsequent resistance. From a moral point of view, the presence of aircraft will cause concern among our adversaries and will be a powerful comfort for our troops.

General Lyautey's political weight was important and the Minister of War, General Pierre Roques, granted his request on April 14, 1916, by sending him two squadrons of eight Voisin aircraft, an aeronautical park, and an aerial photography section, all under the command of Captain Maurice des Prez de la Morlais, who became the head of the aeronautical service in Morocco.

However, there were a number of practical difficulties. The aeronautical fleet landed in Casablanca in June 1916 and the first squadron, numbered 302, was able to reassemble its aircraft. It was immediately sent to Taza to fight the troops of the rebel leader Abd el Malik (grandson of Abd El-Kader, who led the resistance against French colonization in Algeria in the 19th century), who, from Spanish Morocco and with German support, maintained an important rebellion center. Meanwhile, the second squadron, numbered 305, remained without aircraft for a long time and only received its first Farman aircraft in November. It remained in Casablanca to monitor the coastal region. During the winter, two storms destroyed most of the aircraft of both units in January and March 1917. The few surviving aircraft were then used in March to provide air cover for an armed column launched from Fez and Taza against Abd el Malik's troops, whose positions they reconnoitered and bombed their camps. Re-equipped, the 302 remained in the Fez region while the 305 was sent to the Agadir region in April 1917. In June 1917, the airmen achieved a great sporting feat by flying their Farman aircraft from Meknes to Bou Denib, which led them to fly over the Atlas Mountains, which reached 2,200 meters – a great performance for the time !

At this time, General Lyautey was back in Morocco, having left in December 1916 to become the French government's Minister of War, a position he held until March 14, 1917. His return coincided with the arrival of important air reinforcements and a reorganization of military aeronautics. The escadrille 302 became the 551 and, with its two detachments in Fez and Taza, continued to support the operations against Abd el Malek. The 305 became the F 552 and took part in several operations in the interior of the country. The reinforcements allowed the creation of four new escadrilles, F.553 which settled in Meknes in July 1917, followed in September by F.554 in Marrakech, VR 555 in Taza and AR 556 in Rabat, which made it possible to provide air reinforcements to French troops throughout the Moroccan territory, to supply and maintain liaison with isolated outposts and above all, to map these hitherto poorly known regions. The command of the air force in Morocco was given to Major Etienne Cheutin.

Many other battles awaited the Moroccan air force in 1918, which was reinforced by two new squadrons in April (VR 557 at Oujda and VR 558 at Tadla). The operations confirmed the unsuitability of the Voisin bombers for operations and the flexibility of the Farman, a tropicalized version of which was produced. The arrival of the armistice had little impact on the fighting in Morocco, where aviation was to become more powerful with the arrival of the Breguet 14s, and which were to play a decisive role in the Rif war which began in 1921. The last pacification operations were not completed until the early 1930s.

5.3 – Aviation in Tunisia 1914–1918

At the other end of the Maghreb, Tunisia was to experience major fighting that required its military air force, not from natives but from insurgents from outside its borders, in Tripolitania.

This desert region, taken from the Ottoman Empire by Italy in 1912, was not in fact controlled by the latter, which only occupied the cities of Tripoli and Benghazi as well as a thin coastal strip. The interior, entirely desert, remained under the control of indigenous chiefs, including the Senoussi religious brotherhood, which entered into rebellion and inflicted bitter military defeats on the Italian colonial army. Turkish and German agents provided them with discreet assistance and urged them to take military action against French Tunisia, which the Senoussi chiefs refused, prioritizing the fight against occupying Italy. But in 1915, a local leader emerged among the Senoussis, a certain Khalifa Ben Asker, who fought against the Italian occupier. In February 1915, he tried with his men to seize the town of Nalout, but was defeated by a relief expedition organized from Tripoli by the Italian army. He then took refuge with his family in Tunisia where he was interned with his two wives in Kébili by the French authorities. He escaped and resumed his place in the Senoussi rebellion which was growing in strength. With more men, he laid siege once again to the town of Nalout. The Italian commander decided to evacuate the city; he was attacked by Ben Asker and his troops were decimated and captured, although a detachment managed to take refuge in Dehibat in Tunisia.

Nalout being an important locality close to the Tunisian border, the French authorities sent an emissary, Adjudant Ragaru, who was able to meet Khalifa Ben Asker and discovered that he was advised by a certain Sheikh Toueti who is in reality a German agent... Ragaru was able to inquire about the state of the Italian prisoners but the negotiations he undertook with Ben Asker in order to free them came to nothing. He then escaped from the city and returned to Tunisia. The officers of the colonial affairs continue the negotiations by interposed emissaries: Ben Asker tried to negotiate the release of his wives as well as access for the natives of Tripolitania to the Tunisian markets.

Above: Near the high ranking officer in white dress in Captain Maxime de Lafarge, commanding officer of escadrille F.301. The Farman F.410 behind him carries a threat written in Arabic on the front of the fuselage. (Coll Jack Herris)

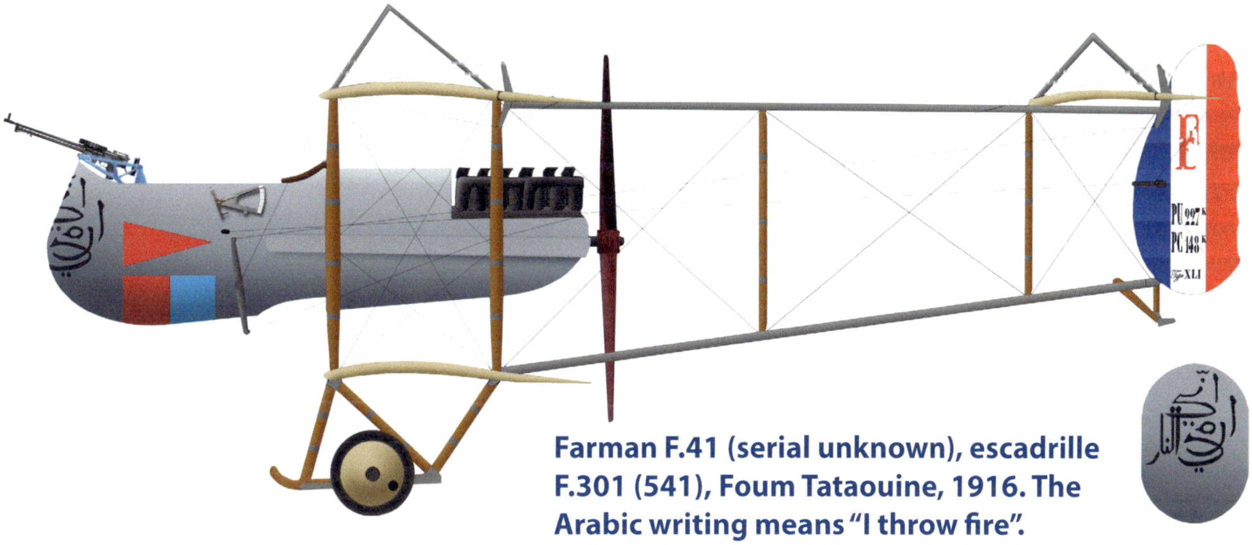

Farman F.41 (serial unknown), escadrille F.301 (541), Foum Tataouine, 1916. The Arabic writing means "I throw fire".

Above: Another F.41 of escadrille F.301, carrying the threat "I throw fire". (Coll Jack Herris)

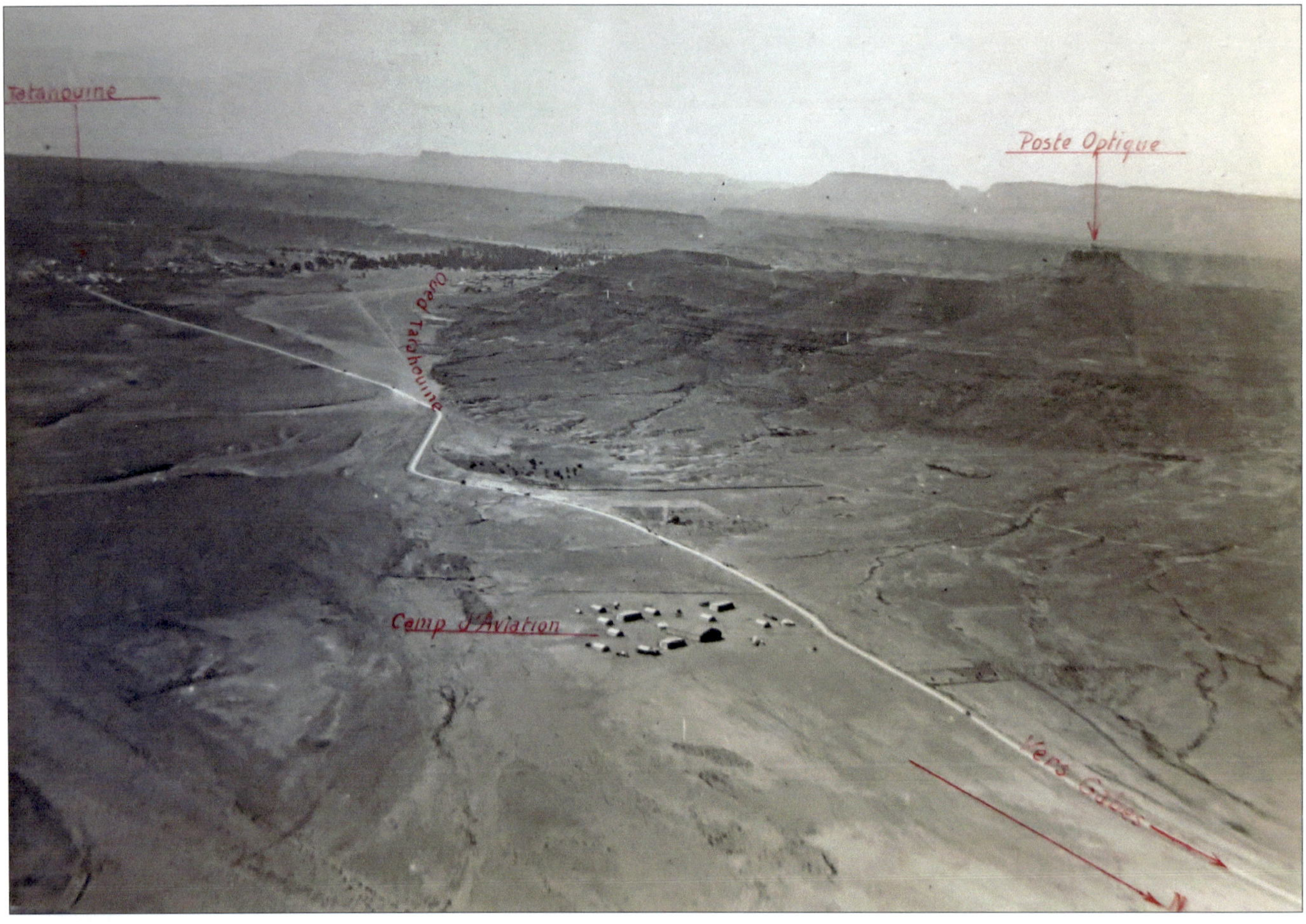

Above: The aviation camp of Foum Tataouine, home of escadrille F.301 in 1916.

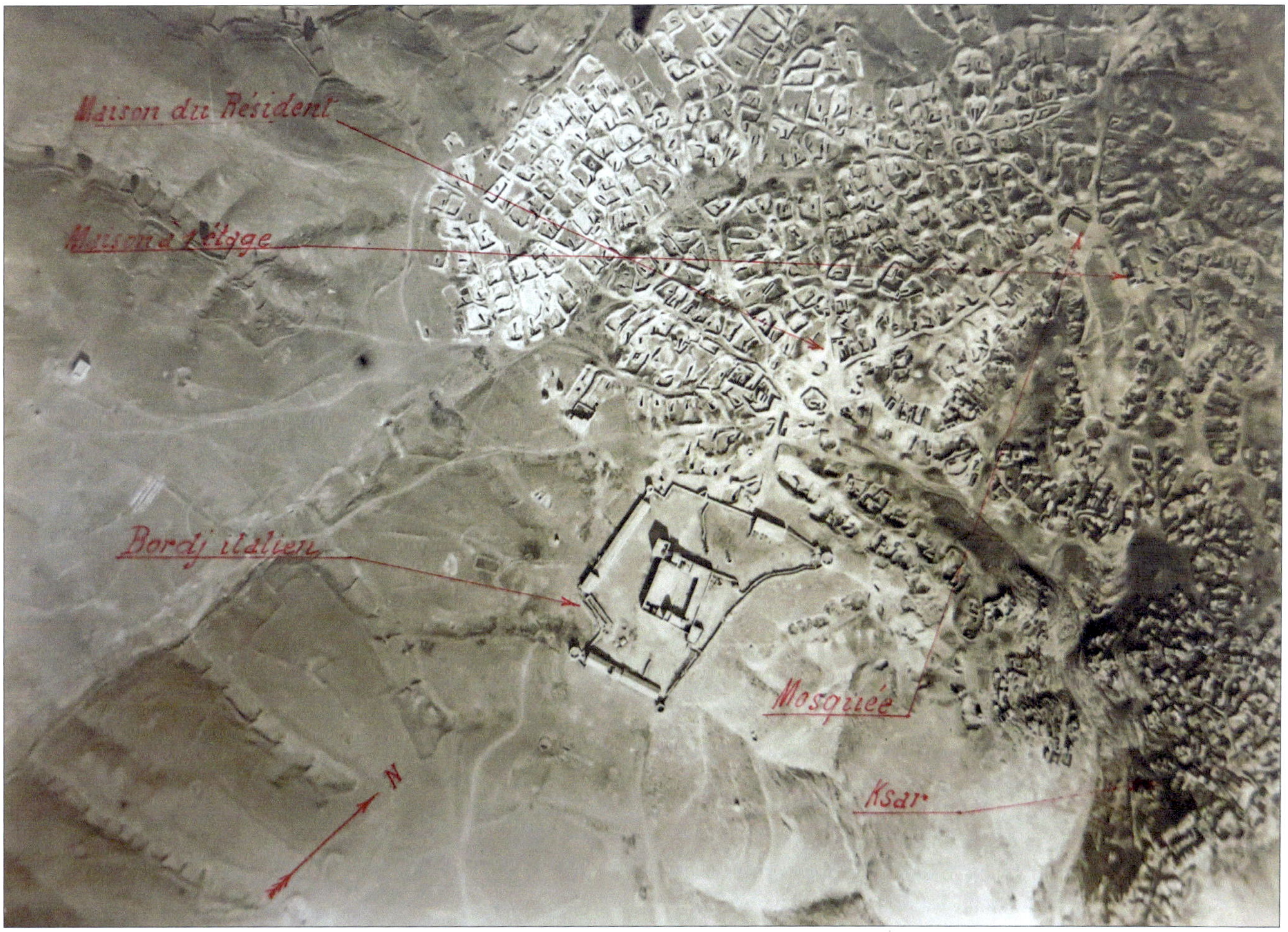

Above: Aerial view of the city of Nalout in Libya, center of Senoussi rebellion, and bombed by escadrille F.301 on September 15, 1916.

Above: Rescue mission in the desert to try to find the missing Farman n°3101 of S/Lt Genet de Chatenay and Lt-Col Henri Le Boeuf after September 15, 1916. The empty plane will be found four months later and their bodies one and two years later, respectively.

Nothing came of it... Further south, the Italian garrison of Ghadames, isolated from the coast by the area controlled by Ben Asker, judged it more prudent to abandon the city and take refuge in French Tunisia after destroying its munitions.

On August 15, 1915, Khalifa Ben Asker decided to attack the French authorities. He issued an ultimatum to the chief of the Dehibat garrison located just across the border and prepared to attack. The Senoussi chiefs tried to dissuade him but he did not care about their advice and launched a few raids against French patrols, killing a few soldiers. This beginning of insurrection made him receive some rallies of natives from the whole region, reinforcing his own troops. On September 25, 1915, he reached a new level by launching an attack on a fortress, that of Retmsa, whose garrison resisted and was rescued by the arrival of a column from Tataouine to help it. On October 1, 1915, the army of Khalifa Ben Asker attacked the post of Oum Souigh where part of the garrison decided to surrender. But a lieutenant and a hundred men decided to resist at all costs and entrenched themselves in a small area. They were finally rescued eight days later by a column of 1,000 men from Tataouine who also re-established the link with the garrison of Dehibat.

The Senoussi rebels returned to Tripolitania... And turned against Egypt by attacking the British troops in November 1915, who inflicted a serious defeat on them at Sollum in January 1916, using self-propelled guns and aircraft. Defeated by this superior mechanical force, they resumed their attacks on Tunisia. Khalifa Ben Asker, whose prestige had declined since the failure of his attacks in the autumn of 1915, attacked the Bir Kecira post without success on 4 March 1916, then, on 18 June 1916, laid siege for seven days to the village of Dehibat, where the garrison resisted as best it could despite the bombardment he undertook thanks to the two captured Italian cannons at his disposal. Ben Asker then attacked the village of Remada on 26 June and succeeded in destroying a column of 180 men launched to its rescue. He attacked another column of 600 men but this one defended itself, forcing him to return to Nalout.

It was at this time that the first French squadron came into action in Tunisia, which radically changed the face of the fighting. Requested on April 11, 1916 by the Tunisian military authorities, a positive response was given and the F.301 squadron was formed in Lyon at the beginning of July 1916 under the command of Captain Maxime de Lafargue, the former head of the 1912 Biskra squadron, who was obviously the most experienced desert pilot in the French military aviation system. His unit consisted of eight pilots in addition to himself and was equipped with Farman F.41s, six of which were scheduled to be on the line and others in reserve. It embarked from Marseille on July 10 and arrived in Bizerte two days later, from where, passing through Tunis, it landed in Gabès on July 21, 1916 and set up its rear base there by erecting its two Bessonneau hangars and its five Adrian huts, then proceeded to reassemble its aircraft. Unfortunately, an accident struck a few days after its arrival, as a Farman carrying two pilots, Sergeant André Bellot and Captain Claude Lamoret, crashed on the runway on July 27, seriously injuring the two aviators who were evacuated to the Tunis hospital where Lamoret died of his injuries on August 10.

Despite this blow, Captain de Lafargue continued his preparations and the assembly of his equipment. The first training flights were launched on August 16 and the barracks were all completed by the 28th. From then on, he set up his operational base in the village of Foum Tataouine, also known simply as Tataouine and which, nearly 60 years later, would inspire the filmmaker Georges Lucas who came to shoot the first episode of his "*Star Wars*" saga in the region. In 1916, it was a small town at the end of the world located 130 kilometers south of Gabes and which had the advantage of being linked to it by a road that was more or less passable, which allowed for regular supplies by car. 150 kilometers south of Tataouine, the border with Tripolitania and the combat zone. De Lafargue organized an aviation camp there with Bessonneau hangars and light tents for the personnel, and on September 2, 1916, the first operational flights could begin.

The aim of these flights was first to carry out photographic reconnaissance of the rebel zone, to draw up maps, and at the same time to initiate psychological action on the local population. The aircraft were all decorated on the fuselage with an imprecation in Arabic intended to impress the insurgents and reassure the tribes that remained loyal to France, such as "I throw fire", "I rumble and throw lightnings like thunder" or "I spread my wings for the glory of France"...

Lightning was indeed going to fall on the troops of Khalifa Ben Asker. The officer in charge of the colonial affairs office in the sector, Lieutenant-Colonel Henri Le Boeuf, asked that the air force strike directly at the center of the rebellion, namely the town of Nalout. In the words of Captain de Lafargue himself, this was a "brutal repression" that used asphyxiating bombs to reach the rebels who had taken refuge in cave dwellings, as the town of Nalout was built partly on a rocky outcrop. While a small fuel, oil, and ammunition depot was set up in Dehibat, a Farman took off at dawn on September 15, 1916 towards this city and landed there, with a pilot-mechanic and a gunsmith on board. Then, four Farmans took off from Foum Tataouine, each armed with a machine gun, six 90mm shells, and asphyxiating bombs, to lead the attack. One of them, piloted by Second Lieutenant Henri Genet de Chatenay, took Lieutenant Colonel Henri Le Boeuf, the instigator of the raid, as an observer. The four aircraft reached Nalout by air and descended to 500 meters to throw their projectiles, although they were under musketry fire. The bombs fell and ravaged the town, setting fire to an ammunition depot and hitting the house of Khelifa Ben Asker; the asphyxiating bombs created a gas cloud in three places in the town. The Farmans came back to Dehibat to refuel in 90 and 155 shells and left for a new bombing. After having carefully noted the result of their attack, they started the return journey but the plane of the

Above: Armored car convoy in the Algerian desert.

Above: Armed cars in front of Farmans from escadrille 541 in Tunisia 1917

Above: Farman at Ghardaïa April 1917

Above: General Nivelle visits Escadrille 547 in 1918

Second Lieutenant Genet de Chatenay lost its way from the formation for an unknown reason and disappeared. The raid was interrupted and search patrols were organized, all the border posts were warned and a Farman that had remained in Tataouine took off towards the south to start searching... They were unsuccessful and it was only four months later, on January 10, 1917, that Farman F.31 n°3101 was found some 150 km west of Nalout, in the dunes north of the village of Al Borma, in the Algerian desert not far from the Tunisian border. The remains of Lt-Col Le Boeuf were found on January 5, 1918 several kilometers west of the aircraft and those of S/Lt Genet de Chatenay even further west on February 17, 1919.

The town of Nalout emerged after the raid... We do not know the number of victims nor the real effects of the gas shells of which it was the only documented proof of use by the French air force during the war; because of the permanent presence of often changing winds it is probable that the gas dissipated very quickly. If Captain de Lafargue considered that the results were considerable, one element is undeniable: there were no more attacks against French garrisons from that day on. Faced with this show of force, many dissident tribes came to ask the colonial authorities for aman, that is, to ask for forgiveness in exchange for their submission.

After the raid, de Lafargue increased the number of flights of his aircraft and defined strict safety instructions, prohibiting solo flights. The air patrols were combined with patrols in vehicles equipped with machine guns that criss-crossed the desert. As early as January 1917, a request for air reinforcements was made and immediately obtained with the arrival of the escadrille 305 (renumbered 542 in June) which settled in Zarzis and carried out its first operational flights in April 1917; It was soon followed by two new units, squadrons 543 and 544, which settled respectively in Gabès and Médenine, allowing for complete air coverage of the border with Tripolitania, while squadron 301, renumbered 541, remained in Tataouine until the end of the war. Khalifa Ben Asker was not to be heard from again, at least on the Tunisian side of the border. Still fighting against the Italians, he was handed over to them by the Senoussi chiefs and ended up being hanged in July 1922.

5.4 – Aviation in Algeria 1914–1918

Algeria, like Morocco and Tunisia, started the Great War "emptied" of its aviation with most of its pilots, mechanics, and equipment sent to fight in France. It was reconstituted in January 1916 from the Biskra airfield where the escadrille F 306 was created on Maurice Farman.

No notable events took place in Algeria during the first two years of the war, until the autumn of 1916 when a violent revolt broke out in the Aurès massif. It was triggered by the generalization of conscription, imposed by the colonial administration to force young indigenous men to serve as soldiers in the French army or as workers in factories. A system of payment of a compensatory tax allowed the wealthiest indigenous peasants to opt out, which the poor peasants of the Aurès massif in eastern Algeria, already victims of land dispossession by the colonial administration, could not afford. During the night of November 11 to 12, 1916, a thousand insurgents attacked the commune of Mac Mahon (today Aïn Touta), killing the local administrator and wounding the sub-prefect of Batna who was on tour there. Other attacks took place in the region to rescue young indigenous conscripts and destroy a few isolated detachments of French troops. The repression was to be particularly fierce and methodical, with the French army initially deploying 6,500 men to protect the settlers and their villages. Then, with the number of men increased to 14,000 at the beginning of 1917, the actual repression began, which largely employed the air force and the escadrille F 306 of Biskra, as well as a new unit, F 310, formed in December 1916 on the same airfield. Their aircraft were used to carry out aerial reconnaissance to guide the French troops reinvesting the Aurès massif and to protect their supply columns which

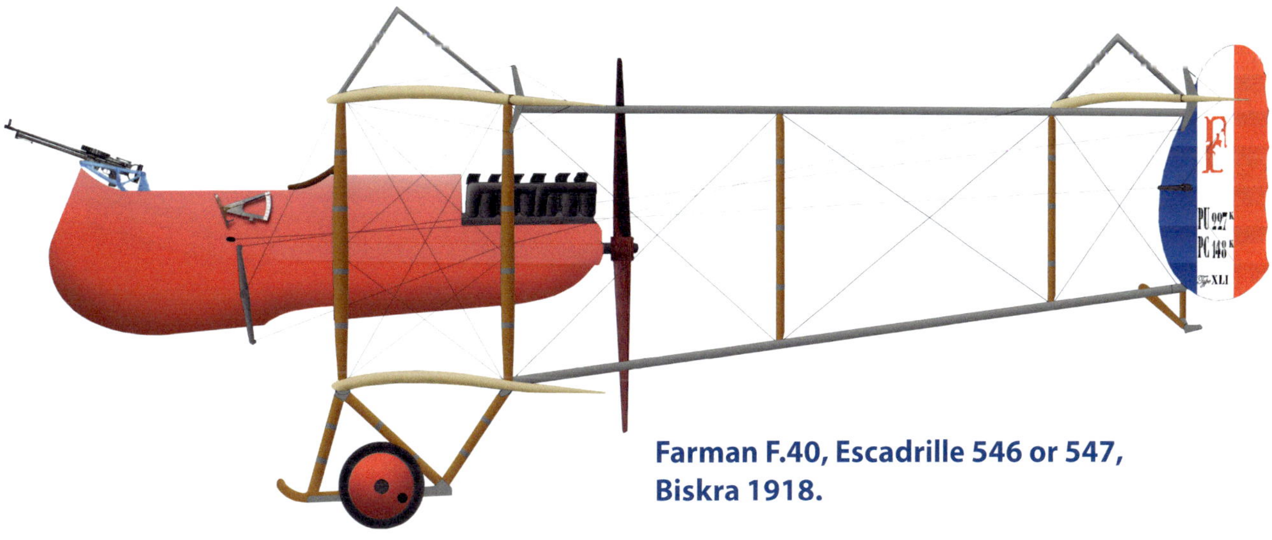

Farman F.40, Escadrille 546 or 547, Biskra 1918.

Above: View of the barracks of the Biskra aviation camp, welcoming the F.306 squadron renamed 547 in June 1917.

Above & Following Page: Autochromes showing the oasis of Biskra (Jules Gervais-Courtellemont)

Above: Farman F.40 of escadrille F.306 in Biskra, April 1917 (Marcel Simian).

Left: Flying above the desert : the common task of pilots from escadrille F.306.

Above: On April 16, 1917, the Farmans from escadrille F.306 carried out a raid from Biskra to Laghouat and Gahardaïa, two oasis in the middle of Sahara.

Above: First step : Laghouat.

Above & Below: Two autochromes showing the oasis of Laghouat (Jules Gervais-Courtellemont)

Above: Each time the Farmans land in the desert, all the local population comes to see those strange birds.

Above & Facing Page: Autochromes showing the oasis of Ghardaïa. (Jules Gervais-Coutellemont)

Above, Below, & Facing Page, Top: The governor of Sahara visits Biskra on June 1917 and inspects the Farmans from escadrilles 546 and 547. (Marcel Simian)

Above: Note the colorful decoration of some Farmans – the one on the left seems painted all red, the one on the right wears color stripes. (Marcel Simian)

Above: Escadrille 547 made a propaganda tour across Algeria during Spring 1917. This Farman is seen here on August 8, 1917, during a stopover in Biskra. (Marcel Simian)

Above: The caption of the photo say, lieutenant Marcel Simian, commanding officer of escadrille 547, stands in front of his "Blue plane", a Farman F.40, in Biskra during the month of November 1917.

Below: On April 1918, General Nivelle (right), former chief of staff of French Army and then military governor of French North Africa, visits the oasis of Biskra and the Farmans of escadrilles 546 and 547. (Marcel Simian)

Above: This photo of General Nivelle's visit is particularly interesting because it allows to compare the difference in shades of the Farmans seen in the background. In the foreground, a particularly decorated Caudron G.4. (Marcel Simian)

Below: Other view of the Caudron G.4 and his very detailed nose art showing a desert owl behind a landscape of the mountains of Algeria. (Collection Pierre Jarrige)

Above: Pilots and observers of escadrilles 546 and 547 of Biskra in 1918. In the center of the photo, wearing crutches, is Captain Laurent, who lost a leg in a flying accident during year 1917. At the right of Captain Laurent is Lieutenant Marcel Simian, commanding officer of escadrille 546.

Below: Pilots of escadrille 547 in 1918. (Marcel Simian)

Above: Old Farmans were still in use in 1919, before being replaced by Breguet 14s. The two Farmans shown there are in Colomb-Béchar in 1919.

Below: Farman F.40, Escadrille 546 or 547, Biskra 1918.

Above: Car convoy in the Algerian desert

Above: Car convoy in the Algerian desert

were always at the mercy of ambushes. They also carried out some bombardments of the rebel villages, intended above all to terrorize them and demonstrate the power of the French army, and they also launched leaflets. The aircraft of the two squadrons operated from Biskra, but also from three country airfields in Batna, Timgad, and Zerbet-el-Oued. Three aircraft were lost due to mechanical problems, but without damage to the crews. At the end of March 1917, the insurgency was defeated at the cost of about 100 combat deaths among the natives who had taken up arms and an unknown number of collateral civilian victims... There were nearly 2000 other natives condemned to various sentences by the military courts.

There would be no more notable revolts in Algeria, except for the one of the populations of the very distant Sahara in the Hoggar, which General Laperrine, recalled from France to find regions that he knew well because he served there from 1902 to 1910, would work to reconquer. The escadrilles F.306 and F.310 provided him with support by flying to the isolated oases of the Sahara while also carrying out propaganda flights over Algeria. The first Saharan raid took place in April 1917 when five Farman planes of escadrille F.310 took off from Biskra to reach the oasis of Laghouat, 290 km to the west, and then joined the one of Ghardaïa, 160 km to the south.

Then, between June 4 and September 16, 1917, the most important raid took place, mobilizing the entire escadrille F.310, renumbered 546, which went on a propaganda tour throughout Algeria to show the French air force to the natives. Under the command of Captain Laurent, the squadron left Biskra for the Algiers region, landing at Boufarik, then flew to the Moroccan border, landing at Tlemcen, from where it flew over several border towns. It then moved towards the Tunisian border via Oran and landed at Bône (today Annaba), then returned to Biskra after several stages, joining the escadrille 547 (former F.306) now commanded by Lieutenant Marcel Simian.

After this memorable raid, the two squadrons were to devote their efforts to the Sahara. A detachment settled in the Ouargla oasis at the beginning of 1918, preceded by a car convoy that had set up a runway and facilities for the Farmans. From Ouargla, a new raid was prepared towards the oasis of In Salah, 600 kilometers away to the south. The preparatory automobile convoy that left to set up the In Salah site was attacked by Tuaregs and massacred 200 km north of their arrival, at the Aïn-Guettara pass on February 1, 1918. A pilot from escadrille 547, Lieutenant René Fondet, was killed in the ambush, as well as nine other members of the squadron, their guide and eight native soldiers. This tragedy did not prevent the field of In Salah from being developed, and then from being reached by four Farmans commanded by Lieutenant Bernard (three aircraft from escadrille 546 and one from 547) on March 14, 1918, making in one day by air a connection that usually takes two weeks by camel and also making the first postal connection to the Sahara because the Farmans also carried mail.

This raid, the first of its kind, prefigured the great aerial crossings of the African continent carried out in the early 1920s. It had a great impact in Algeria and the news even reached France since the press announced it in its inside pages. The desert aviators then had the honor of transporting General Nivelle, appointed Commander-in-Chief of the French forces in North Africa, who carried out an inspection in Ouargla – he returned by air to Biskra on April 24, 1918.

After this date, there were no more large-scale raids, both because of the arrival of summer in the Sahara, where temperatures were extreme, and because of the wear and tear on the equipment, as the old Farman F.40s were at the end of their useful life – the Caudron G.4s delivered to Biskra were not very well suited to the Sahara because of their rotary engines, which did not stand up well to the sand. Two other Algerian squadrons were created in 1918, the 548 in Oran and the 549 in Constantine, without any notable activity.

5.5 – Watching the Waves: Maritime Aviation in North Africa 1914–1918

While military airmen crisscrossed the continent and fought colonial revolts, another front opened up in 1916 on the shores of Algeria, Tunisia, and Morocco: the war against U-boats.

What the surface ships *Groeben* and *Breslau* had not been able to achieve in 1914, i.e. the disruption of maritime traffic from North Africa, the submarines of the German navy were to achieve in 1915 by sinking numerous commercial ships which were generally defenseless. The French and Allied navies did not have the means to provide escort ships in quantity, but the use of seaplanes for coastal surveillance quickly became a solution to be developed.

In April 1916, a Maritime Aviation Center (CAM) was created in Bizerte, on the main French naval base in Tunisia, using FBA type C and receiving the first Donnet-Denhauts with Hispano 150 hp engines. The seaplanes, which also used battle stations and rest stations in Bône, Kélibia, Tabarka, Galite Island, but also Marsala in Sicily and Cagliari in Sardinia, had no success in 1916 and did not spot any submarines.

In 1917, the construction of a network of Maritime Aviation Centres along the entire coastline of French North Africa began, allowing for a constant surveillance of its coastline. The first of these, the Algiers CAM, became operational in March 1917 with Donnet-Denhaut 150 hp Hispano seaplanes.

However, in a hurry and before most of the CAMs were built, the naval staff sent two seaplane carriers to North Africa, the *Dorade*, which operated in Casablanca from April to June 1917 with three 130 hp Clerget (type C) seaplanes with folding wings, and the *Normandie*, which operated off the coast of Algeria in July 1917 with four other similar 130 hp FBAs. The experiment was quickly interrupted due to the lack of power of these seaplanes, which were too light to carry a significant offensive load, as well as the difficulty of launching and recovering them in heavy weather.

Above: Seaplane Donnet-Denhaut 200 hp Hispano from CAM Casablanca, marked with the identification letter "C" given to this Center. (Collection ARDHAN)

Right: His Donnet-Denhaut seaplane, from CAM Oran, shows a rare personal decoration (a moon crescent) and a stylized identification letter. CAM Oran was suppressed in early 1918 and became a Combat Post from CAM Arzew. (Collection ARDHAN)

Above: Aerial view of CAM Arzew in 1918. (Collection ARDHAN)

Above: Closeup of the Bessonneaux hangars from CAM Arzew. (Collection ARDHAN)

Above: A Donnet-Denhaut 200 hp Hispano of CAM Arzew. The seaplanes of this Center had "W" for identification letter. (Collection ARDHAN)

Below: Donnet-Denhaut 200 hp I lispano from CAM Arzew taxiing on the water. This seaplane show a rare personal decoration, the name "*Netto*". (Collection ARDHAN)

Above: The U-Boot U 39 is shown here in the bay of Kotor (Montenegro). Two German submarine were neutralized by French maritime aviation during WW1 and U-39 was one of them, damaged after bombings from seaplanes of CAM Arzew on May 18, 1918. (Collection Edwin Sieche)

Left: Same U-Boot, which took refuge in the Spanish port of Cartagena where it remained interned for the rest of the war.

Above: Ensign Joseph Leray, winner of the U-39.

In April 1917, CAM Oran was set up, followed by CAM Sousse in Tunisia in July 1917, Bône (Annaba) in September 1917, and then in October CAM Marsala in Sicily, with the agreement of the Italian government. In January 1918, CAM Casablanca in Morocco came into action, followed by CAM Arzew (replacing the Oran one) in Algeria in April 1918, and finally CAM Djidjili in August 1918. Two coastal landplane squadrons were created in August 1918, the G.489 at Mazagan in Morocco on Caudron G.4 and Voisin VIII, and the G.490 on Caudron G.4 at Sfax in Tunisia. Airships also operated with seaplanes, and were based in three centers: in CAM Bizerte (Sidi Ahmed), where two airships operated at the end of 1916; the Baraki Center near Algiers, whose two airships began their patrols at the end of 1917; and the Oran-la-Sénia Center, whose two airships were operational in February 1918.

The equipment used by all the Maritime Aviation Centers in Africa was the Donnet-Denhaut seaplane in its various versions, first with a Hispano 150 hp and Lorraine 160 hp engine in 1917, then with a Hispano 200 hp engine in 1918. There were also Tellier seaplanes with a Hispano 200 hp engine at Algiers, Bizerte, and Sousse, as well as its version equipped with a 37 mm gun at the Algiers, Arzew, Bizerte,

Below: Damaged Donnet-Denhaut 160 hp Canton-Unné from CAM Bizerte, in 1916. This Center was the first to be operational in north Africa, In April 1916. In this time, the seaplanes had no identification letters.

Above: Autochrome showing the city of Sousse in 1931 – and the port were the seaplanes of CAM Sousse were stationed, in Bessonneaux hangars. (Frédéric Gadmer)

Marsala, and Sousse naval bases. CAM Marsala also received Georges Lévy seaplanes with Renault 280 hp engines in 1918.

The first result came on April 29, 1917 when a U-Boat was attacked by a patrol of aircraft from CAM Bizerte. As always for seaplane attacks against submarines during WWI, these were bombs dropped on a U-Boat wake which went into a dive, the results of which are unknown, not to say very unlikely. But the main result was to put the submarines in a state of insecurity, since until 1916 they sailed on the surface and sank their target with cannons, diving only in case of danger… In 1917, merchant ships started to sail in naval convoys and the seaplanes of the North African CAMs were more and more solicited for escort missions, rather than for patrols. On May 7, 1917, an attack took place which was considered credible enough to be covered by the media: a submarine was spotted around 7 a.m. by the aircraft piloted by Enseigne de Vaisseau de 1ere classe Auguste Biroard, who warned the Algiers center from where two Donnet-Denhaut 150 hp Hispano seaplanes (n°261 coded A7 of Corporal Henri Nègre and Quartermaster Médrard Capdegrelle, as well as the n°251 coded A8 of Sous-Lieutenant Léopold Coren and Ensign first class Charles Darchis de Lantier). They found the enemy submarine in the reported area and Coren's aircraft bombed it while it was diving and damaged it because its wake left air bubbles and a long trail of oil… Corporal Nègre's A7 launched its bombs in turn on the wake. All the aviators were decorated and their exploit was even the subject of an article published on September 13, 1917 in the magazine "*La guerre aérienne illustrée*"… Coren, when questioned, declared to the press that "*at the minute of our success, I thought intensely about the sixty or so poor fools who were irresponsible for the gesture that mobilized them and who were going to pay with their lives for the shame of serving the nation that had started the war.* Nevertheless, the German and Austro-Hungarian archives are formal: no U-Boat would be lost this month in this sector of the Mediterranean!

There were a total of 22 U-Boat attacks carried out during the war by the African and Portuguese CAMs, of which 7 were carried out by CAM Bizerte (including one attack by a naval airship), 6 by CAM Oran or its successor CAM Arzew, 2 attacks for each of the Algiers, Bône, Sousse, and Aveiro CAMs, and one by CAM Marsala in Italy. Only one of these attacks was proven to have neutralized an enemy submarine, carried out by CAM Arzew on 18 May 1918.

Above: Donnet-Denhaut 200 hp Hispano from CAM Sousse in 1917. This Center had "S" as identification letter. (Collection ARDHAN)

Below: Another Donnet-Denhaut 200 hp Hispano from CAM Sousse on the water. (Collection ARDHAN)

On May 17, 1918, a naval convoy in the vicinity of the Strait of Gibraltar was attacked by two U-boats, the U 50 accompanied by the U 39, which was one of the most formidable submarines, since the three captains who commanded it had sunk no less than 157 ships in the Mediterranean, i.e., the second most important hunting record of the war. A radio call was made from the convoy, which was 60 miles west-northwest of Oran, and relayed to the commanders of the local Maritime Aviation Center. As another convoy bound for Bizerte was expected the next day off Oran, CAM Arzew sent to meet it the next day, May 18, a section of two Donnet-Denhaut seaplanes (probably equipped with hispano 200 hp engines), commanded by the observer of one of the aircraft, Enseigne de Vaisseau de 1ere classe Campardon. They found the convoy 75 miles east-northeast of Oran, which they escorted for a few moments before taking the way back out of fuel. On their way back, they met the U 39 of Kapitanleutnant Heinrich Metzger in a semi-dive 20 miles from the convoy, heading northwest. The two seaplanes launched an attack and dropped their bombs in the wake of the submarine which dived, without causing any damage.

Without radio, the French airmen could not give the alert until they returned to Arzew at 8.45 am. Another section of two other Donnet-Denhauts was about to take off on their arrival, commanded by one of the pilots, Ensign Joseph Leray. But they had to delay their departure because of bad weather and could not join the convoy until 12:00 p.m., which they escorted for half an hour before making their way back to the southwest. They then saw the submarine sailing on the surface towards the north-west (3636 North, 0002 West), and, playing with the clouds, were able to approach it discreetly and surprise it before it had time to dive. The German crew had indeed seen the attack and was at a depth of 12 meters when two bombs, thrown by the crew composed of Ensign Joseph Leray and Quartier Maître Henri Baconin, exploded near the hull and opened a waterway flooding a compartment.

Leray saw only an oil stain and floating debris. He landed to recover some but was surprised by the submarine surfacing, having to take off in a hurry surrounded by enemy fire... Also without a radio, the two Donnet-Denhauts could not reveal their odyssey until they returned at 2 pm. A third section of two seaplanes led by Ensign Lassale left at 2:25 pm to reach the damaged submarine at 3:30 pm. But if the submarine could no longer dive, its anti-aircraft gun was still operational and fired on the French aircraft, which had to take height to hide in the clouds during their approach. They could only carry out a disorganized attack which missed the U-Boat, even though it had lost two servants of the artillery piece, who fell overboard during the panic of the bombardment.

The damaged submarine escaped from the other patrols and surface units, which had been warned late by the Arzew center, which did not have a radio. With the help of another U-Boat, it managed to reach the Spanish port of Cartagena around 11 pm. The French journalists present in this neutral port were able to see the damage to the hull, the periscope, and the deck, and to report that its commander offered tea to German personalities in the city... Too damaged to be repaired, the submarine and its crew were interned by Spain for the rest of the war.

This naval victory was widely publicized, since on July 31, 1918, an article in the *Figaro* newspaper stated that aircraft were responsible for the victory over the U39. All the participants in the attack received a citation and the winning crew, Ensign Joseph Leray and Quartier Maître Henri Baconin, were respectively awarded the Legion of Honor and the Military Medal. They had the honor of being treated like the great fighter aces because their names were mentioned in a newspaper... Leray took advantage of his laurels to ask to be trained on a fighter plane in order to be able to fight the German air force, which was granted to him and he obtained his transfer to the fighter squadron of the CAM Dunkirk just before the armistice.

5.6– Portugal, Senegal, Indochina – Air Units from the Ends of the Earth

To complete this exhaustive view of the French aviation on all fronts, it is necessary to mention the fate of the French air formations that were created during the conflict in Portugal, Senegal, and Indochina.

5.6.1 – In Portugal

Portugal was a neutral country at the beginning of the First World War, bound by an old alliance treaty with Great Britain, which did not, however, lead the government to enter the conflict, despite merchant ships being sunk by U-boats. A violent military coup on May 14, 1915 brought to power the Republicans who, under pressure from the British government, decided on February 23, 1916 to seize all German ships that had taken refuge in Portuguese ports. As a result, on March 9, 1916, Germany declared war and plunged the country into the conflict. The Portuguese army was to actively fight against the German colonial troops of General Von Lettow-Vorbeck in Mozambique and organize a small expeditionary corps fighting in France in 1917. The Portuguese Navy had to face the U-boats in the Atlantic and the French Navy came to its aid by proposing the installation of a Maritime Aviation Center which was set up in Sao Jacintho, near the port of Aveiro. The first construction works started in January 1918 and the first seaplane patrols started the following May, on Donnet-Denhaut 200 hp Hispano and Tellier 200 hp Hispano as well as Geroges Lévy 280 hp Renault delivered from September 1918. The French airmen, who suffered no losses, carried out five attacks against U-boats in September 1918, two of which are described in the archives. The center was then dissolved after the armistice and its aircraft were handed over to the Portuguese Navy.

5.6.2 – In Black Africa

The French colonial empire was not limited to North Africa and also extended south of the Sahara, firstly in Black Africa

Above: We're closing ! This is what says the sign held by this sailor from the CAM Sousse, on the day of the armistice of November 11, 1918. Note the personal decorations cut from the hulls of the Center's seaplanes held by two other sailors. (Collection ARDHAN)

Donnet-Denhaut 150 hp Hispano n°251 « *Lou Gobi* » of CAM Alger, 1917. This profile was drawn according to a diagram showing the decoration of the seaplane use by Sous-Lieutenant Léopold Coren to bomb and claim a U-Boot on May 7, 1917. The U-Boot, confirmed to Coren, did in fact escape the attack.

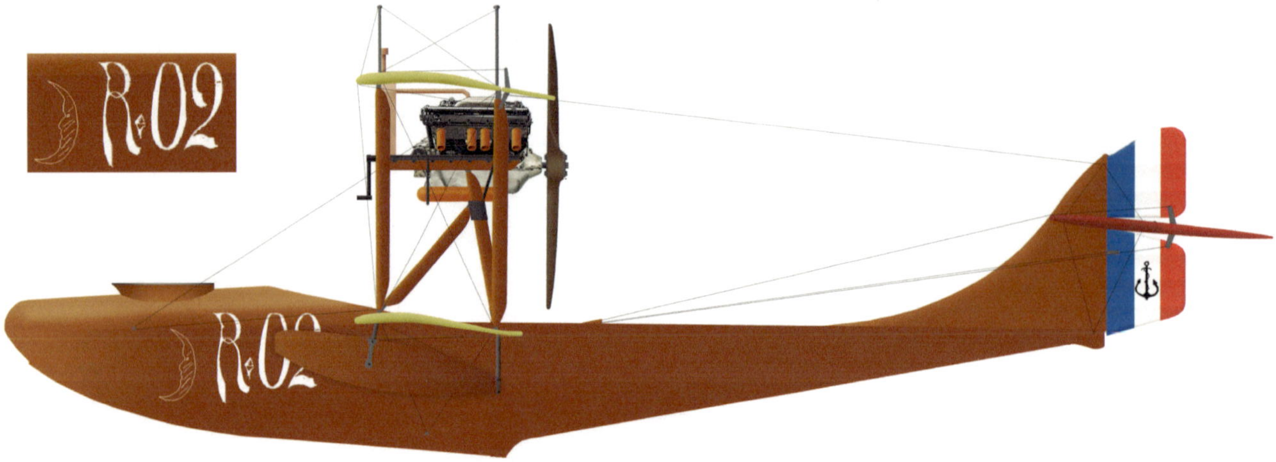

Donnet-Denhaut 150 hp Hispano n°281 of CAM Oran, 1917.

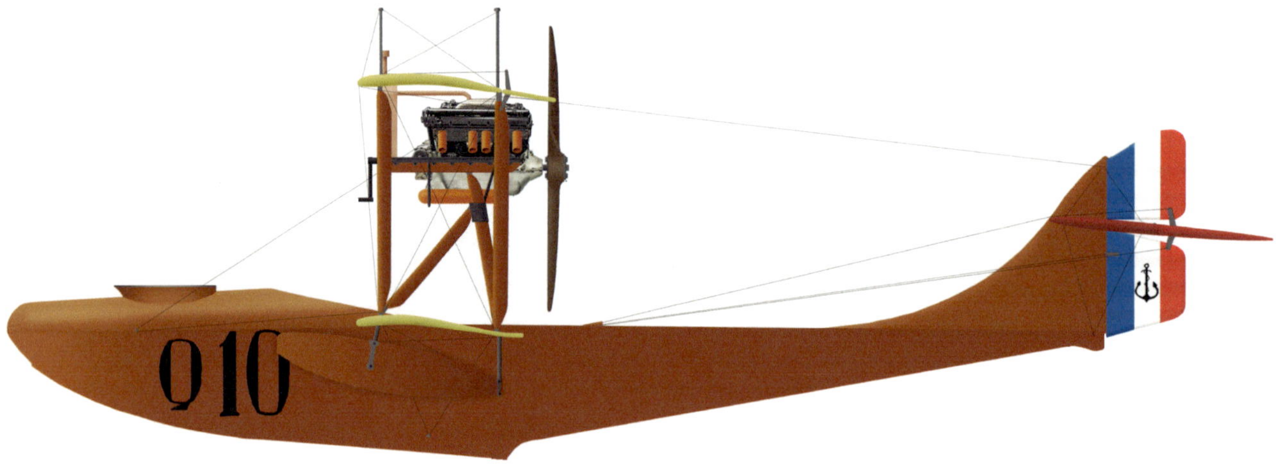

Donnet-Denhaut 150 hp Hispano (serial unknown) of CAM Bone, 1917.

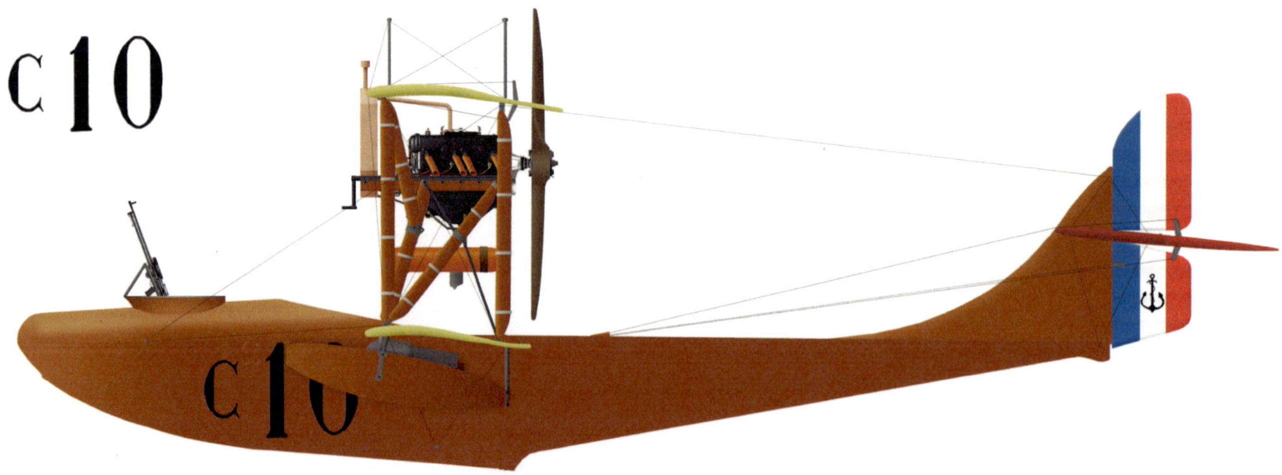

Donnet-Denhaut 200 hp Hispano n°1271 of CAM Casablanca, 1918.

Donnet-Denhaut 200 hp Hispano n°1525 of CAM Arzew, 1918. The seaplanes who attacked U-39 on May 18, 1918 are not precisely known.

Tellier 200 hp Hispano n°27 of CAM Bizerte, Early 1918. In this time, the Center had "H" as identification letter.

Tellier 200 hp Hispano n°124 of CAM Bizerte, mid 1918. In this time, the identification letter of the Center switched to "Z".

Above: Donnet-Denhaut 200 hp from CAM Aveiro (Portugal) in 1918. This seaplane carries the nickname "Patte folle" (Crazy leg).

by grouping together Mauritania, Senegal, Guinea, Niger, Mali, Upper Volta (now Burkina Faso), the Ivory Coast, and Dahomey, forming the French Oriental Africa (A.O.F.). In the center of Africa, France occupied Chad, Oubangui Chari (present-day Central Africa), Gabon, and Congo, forming French Equatorial Africa (A.E.F.). To the east of the African continent, France also possessed the territory of Djibouti and the island of Madagascar. In this vast colonial empire, aviation aroused a certain interest among colonial governors who perceived the potential for exploration and liaison that aircraft represented in countries without means of communication.

William Merlaud-Ponty, Governor General of the A.O.F., was the first to make this decision and on November 28, 1910 he obtained from the Minister of the Colonies the purchase of four airplanes, two Farman biplanes and two Nieuport monoplanes. The purchase contracts were ratified on December 28, 1910 for 117,425.10 francs (60,172.55 F for Farman and 57,252.55 F for Nieuport), including with the delivery of the aircraft to Bordeaux the training of one pilot per aircraft. The two Farman, accompanied by two Bessonneau hangars (bought for 12 591.30 F), arrived in the port of Dakar on April 3, 1911. Even before their arrival, on March 17, a detachment of 20 riflemen began to set up an airfield near the village of Bambey, located 80 km east of Dakar and served by the railway under construction. The new head of the Senegalese air force, Lieutenant Marcel Sido, was

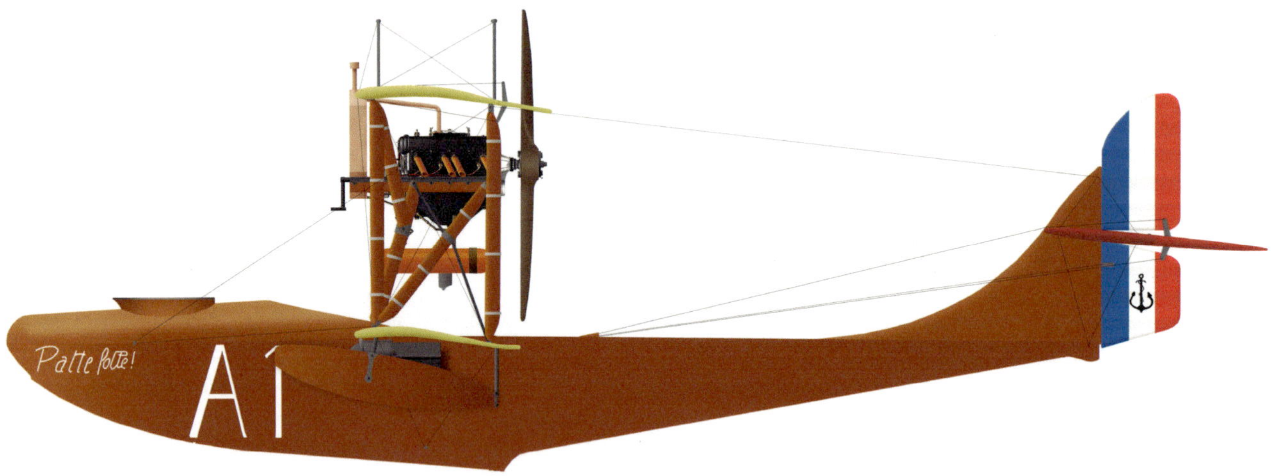

Donnet-Denhaut 200 hp Hispano (serial unknown), CAM Aveiro, 1918.

Above: Nieuport monoplane flown by lieutenant Albert Féquant in 1911 in Sénégal.

the only pilot for the moment, along with the young colonial administrator Fernand Carles, who had just been certified as a pilot on February 3. Sido had to supervise the development of the airfield, the work of which was dragging on: all the cut trees had to be removed and the termite mounds had to be treated as obstacles for his fragile machines. It was finally on June 13, 1911 that Marcel Sido was able to make the first flight and rise to fifty meters on one of the Farman biplanes. On November 22, 1911, they were joined by Lieutenant Philippe Féquant who landed in Dakar with the two Nieuport monoplanes. The small group of aviators flew the skies of Senegal on their aircraft, generating great enthusiasm among the local population. On February 10, 1912, Carles on a Farman and Féquant on a Nieuport flew together over the city of Dakar and its surroundings. On May 29, 1912, Lieutenant Féquant took off from Bambey at 5:45 a.m. on one of the Nieuports and landed after a 105 km flight at Ndande, a village between Dakar and Saint-Louis du Sénégal. From there he took off the next day for Louga, 50 km further north, then on May 31 he arrived triumphantly in Saint-Louis du Sénégal, flying over the city and its mosques, the bridge and the Senegal River. All the local population can come and admire his aircraft closely in an aviation festival which takes place on June 2, Féquant giving improvised conferences on the power of military aviation. This may be a bit of a mouthful, as nature quickly got the better of the fragile machines, which were all destroyed by a tornado at the end of 1912.

A similar experiment took place in Madagascar on the initiative of Governor General Albert Picquié, who in 1910 invited one of his subordinates, the administrator Jean Raoult, who was on leave in France, to learn to fly. He did so... and came back to Madagascar with two Blériot aircraft. They were installed in Androhibe, 6 km north of Tananarive, the capital of the island, which Raoult flew over on July 9, 1911. But these fragile machines, bought without spare parts, were quickly grounded because of the combination of the inexperience of their pilot who breaks them on landing and their lack of spare parts.

There was no aircraft in flight condition in the whole of French Black Africa when the war broke out and it was without aviation that the French colonial troops participated with the British in the conquest of the German African colonies, Togo, which fell in 1914, and Cameroon, whose capital Yaoundé was conquered in January 1916.

It was not until 1918 that new aircraft were flown over Senegal with the opening of a Maritime Aviation Center intended to protect the area around the port of Dakar from possible German U-boats. Acting on December 21, 1917 by a decision of the Ministry of the Navy, CAM Dakar saw its aircraft (Donnet-Denhaut 200 hp Hispano) delivered in April 1918 and the first test flights took place in May. However, it is only with the arrival of the pilots that the activity of the CAM could really start in July 1918, without being very active as the weather conditions made the maintenance of the aircraft (reinforced with some Georges Lévy 280 hp Renault) problematic. However, they did not encounter any German submarines and did not suffer any accidents until the armistice.

5.6.3 – In Indochina

The most distant of the French colonies, Indochina, which includes present-day Vietnam, Cambodia, and Laos, also had some aerial experiences before the war. At the beginning of 1910, an Italian garage owner from Saigon, a certain Mr. Vincenzo Ippolito, bought a Blériot XI in France and reassembled it in his workshop in order to have it flown by the city's aviation enthusiasts... Who broke the aircraft on its first take-off, on May 21, 1910. It was a Belgian pilot, Charles Van Den Born, who made the first flight on his Farman on December 10, 1910 in Saigon, in front of the entire city, which celebrated the aviator. Other pilots from Europe followed, including Marc Pourpe who met the man who would become his mechanic, Raoul Lufbéry, a future ace fighter pilot. However, these were only private initiatives. The Governor General of Indochina, Mr. Albert Sarraut, was seriously interested in the question and wished to set up an experimental airfield in Hanoi, in the north of the colony, where the General Government was based. However, in France, the Minister of War and the Minister of the Colonies opted to establish an airfield in Saigon – in the south of Indochina. This disagreement condemned any project that the general government of Indochina, for budgetary reasons, could not carry out alone. Albert Sarraut was recalled to France in December 1913, but he returned to Indochina in the same position during the First World War, in November 1916, and took up the matter where he had left off.

In December 1916, he began negotiations with General Hubert Lyautey, who was then appointed Minister of War, for the creation of an Indochina Air Force. He received a polite refusal because France at war had more urgent concerns... The Ministry of War finally agreed, but on condition that the Indochina colony finance the entire operation. Albert Sarraut accepted... Without really accepting, both financially and because the air force depended on the military hierarchy. This will not be without consequences on the aviation of Indochina during the war.

On January 10, 1917, following the agreement, the General Staff decided to create escadrille V.309 of Indochina, which was more like a half-squadron, with a planned strength of six planes and four pilots. Its purpose was to monitor the border with China, and for this purpose Sarraut ordered the construction of a site at Vi Thuy, near Tong, about 40 km northwest of Hanoi, where a fuel depot was set up. The personnel arrived from France in the port of Haïphong on July 9, 1917, and four days later Sarraut signed a decree officially creating the Indochina aviation service, which he placed under the orders of his general government, and this led to strong protests from the army! Sarraut also reviewed the budgetary claims of the military and had the budget halved. The Indochina air force was then entrusted to the head of the local geographic service, Captain Henri Salel, who was not a pilot.

Escadrille V.309, which became the V.571 in June (and was designated locally as the escadrille No. 1), had only three pilots in its ranks: its leader, Second Lieutenant Maurice Maldidier, Second Lieutenant Charley Ricou, and Second Lieutenant Henry Wintrebert, a colonial administrator in civilian life. The first two were veterans of the Orient Air Force and had just returned from convalescence, to the point that Maldidier was in the process of being struck off the aircrew list – that is to say that the army had scraped the bottom of its drawers for Indochina... And this was also the case for the aircraft, eight Voisin LAS from the Orient Air Force, where they had been discharged the previous year and spent a year in a depot! Four of them were landed in the port of Haiphong in August 1917 and reassembled by the mechanics who could only note their deplorable condition. Second Lieutenant Maldidier made the unit's first flight on one of them on August 17, flying over the Red River Delta, then on August 31 Ricou made a test and reconnaissance flight of 170 km in 1h38.

It was then that the prisoners of the Thai Nguyen penitentiary, some sixty kilometers north of Hanoi, insurgently took over the village after having killed their guards. The village was immediately reconquered by the French army, but the prisoners withdrew into the hills to wage guerrilla warfare. The head of the French detachment, Colonel Berger, called upon the aviation who made two reconnaissance flights on September 30 and 31, 1917, in an aircraft belonging to S/Lt Ricou. It did not bring much because of the fog which prevented it from communicating its observations to the ground. To make matters worse for the squadron, S/Lt Maldidier, who fell ill, died in Hanoi hospital on September 18. Ricou took over the command of the squadron but only commanded himself because Henry Wintrebert was recalled by the colonial administration. The army sent only two pilots as reinforcements: Sergeant Louis Chaffanjon in March 1918, and Corporal Cao-Dac Minh (an indigenous pilot), the following month. The Voisins were out of breath and did not fly much. In April 1918, Lt Ricou was ordered to fly one of his aircraft during a military review in honor of the Annam emperor Khai Dinh. Judging his aircraft too dangerous, he refused. He was then dismissed from his command and sent back to the army. Louis Chaffanjon, appointed second lieutenant, took command... and was injured at the beginning of May 1918 while trying to take off the last available Voisin. Escadrille V.571, without any aircraft, did not receive any others despite the negotiations undertaken by Albert Sarraut to have Nieuports delivered by the army, but which he refused to pay. It did not fly anymore before the armistice and its only remaining pilot, Corporal Cao-Dac Minh, was sent to Saigon where a second air unit had just been created.

This second squadron was unsuccessful as the first one. Created on April 6, 1918 by a decree of Albert Sarraut, it had only two pilots: its leader, Captain Jules Condamy, and Sergeant André Rusch, whom Cao-Dac Minh joined. Settling on the Phu Tho airfield, which is the racetrack of the city of Cholon, west of Saigon, it received the four Voisin LAS stored in Saigon plus two others delivered from France in July 1918. Unfortunately, they were not in better condition than those received by escadrille No. 1, and one of them had missing parts and could not be reassembled... A hangar was built with difficulty and it was not until October 1918 that the first flights could take place – Cao-Dac Minh, promoted

Above: At the right of the photo, wearing a white colonial hat, general Riou, new Governor of Madagascar, talks with aviator and colonial administrator Jean Raoult, on July 7, 1911, in Androhibe airfield. Raoul did break the Bleriot on his first landing… And did the same for the second Bleriot on July 28, ending the short-lived colonial aviation in Madagascar.

to sergeant, broke his aircraft on his first flight where he was injured when he crashed in the rice fields. He was killed on December 27, 1918 while attempting a flight over Saigon on another Voisin. All Voisin planes were banned from flying and their Canton-Unné engines (seven of which survived) were used for gliding boats. Because of the lack of suitable equipment and pilots, Indochina aviation only made a few flights during the war and it was only in 1920, with the delivery of the Breguet 14s from France, that it could really take off.

10 worn-out Voisin LAS from the Air Depots of Armée d'Orient: that's all Indochina received during the war. The two little escadrilles formed, one near Hanoï and one in Saigon (today Hô Chi Minh City), could make only very few flights.

French Losses in Africa 1914–1918

Date	Rank	Pilot or Gunner, Fate	Unit	Plane Piloted
6/30/16	EV1	Goumaud, André (pilot - KIAcc)	CAM Bizerte	FBA 100 hp Type C n°135 "B.2"
7/2716	Sgt Cne	Bellot, André (Pilot - WIAcc) Lamoret, Claude (Pilot - WIAcc, DOW 10/08/1916)	301 (541) 301 (541)	Farman F.40
9/15/16	Lt Lt-Col	Genet de Chatenay, Henri (Pilot - MIA) Le Bœuf, Henri (Observer - MIA)	301 (541) Colonial affairs	Farman F.41 n°3101
9/29/16	Lt Sol	Bourges (Pilot - OK) Daumas, Pascal (Observer - KIAcc)	301 (541) 301 (541)	Farman F.40
12/13/16	QM Mot	Gille, Henri (pilot - MIA) Raoul, Daniel (Observer - MIA)	CAM Bizerte CAM Bizerte	Donnet-Denhaut 150 hp n°61 "B.4"
1/26/17	MdL Mot	Bey, François (Pilot- KIAcc) Loir, Henri (Observer - KIAcc)	CAM Bizerte CAM Bizerte	Donnet-Denhaut 150 hp n°66
2/26/17	Mot	Parrimond, Baptistin (Observer - KIAcc)	CAM Bizerte	
5/7/17	S/Lt	Le Borgne de Kerambrosquer, Henri (Pilot - KIAcc)	CAM Bizerte	Donnet-Denhaut 150 hp n°242 "B.47"
6/25/17	Lt Sol Lt	Poggi, Maurice (Pilot KIAcc) Vitalis, Victor (Pilot - KIAcc) Claude (Observer - KIAcc)	302 (551) 302 (551) 302 (551)	?
7/31/17	QM	Garnier, Michel (Pilot - KIAcc)	CAM Bizerte	
8/17/17	Mtre QM	Guiliani, Joseph (Pilot - KIAcc) Jaffro, François (Observer - KIAcc)	CAM Bizerte CAM Bizerte	Donnet-Denhaut 150 hp n°242 "H.2"
8/19/17	Mot	Vatine, Gaston (Observer - KIAcc)	CAM Bizerte	Donnet-Denhaut 150 hp n°249
9/4/17	MdL	Prom, Lionel (Pilot - KIAcc ?)	551	?
9/26/17	Mot QM	Chirpaz, Georges (Pilot - MIA) Geay, Jean (Observer - MIA)	CAM Oran CAM Oran	Donnet-Denhaut 150 hp n°280
10/10/17	MdL MdL	Pesché André (Pilot - KIAcc) Escourrou, Armand (Pilot - KIAcc)	546 or 547 546 or 547	Farman F.40
11/8/17	SM QM	Le Gall, Louis (Pilot - MIA) Salmon, Ange (Observer - MIA)	CAM Bône CAM Bône	Donnet-Denhaut 150 hp "Q.1"
1/17/18	Lt	Bourkadam, Constant (Pilot - KIAcc)	556	Nieuport
2/1/18	Lt MdL	Fondet, René (Pilot - KIA) Hours, Armand (Gunner - KIA)	557 557	On the ground
6/20/18	QM QM	Savary, Paul (Pilot - MIA) Riche, Gaston (Observer - MIA)	CAM Marsala CAM Marsala	Donnet-Denhaut 200 hp n°1136
8/19/18	QM EV 1	Faveau, Gaston (Pilot, KIAcc) Poisson, Jean (Observer, KIAcc)	CAM Sousse CAM Sousse	Donnet-Denhaut 200 hp
9/19/18	Sol	Geille (Observer - KIAcc ?)	?	?
10/6/18	SM	Reguer, François (Pilot - MIA)	CAM Alger	Donnet-Denhaut 200 hp
10/13/18	SM Mot	Blanchard, Ferdinand (Pilot - KIAcc) Rastouil, Guillaume (Observer - KIAcc)	CAM Bizerte CAM Bizerte	Donnet-Denhaut 200 hp n°1226
10/21/18	Cne EV 1	Ducasse, Marcel (Pilot - KIAcc) Marre, Georges (Pilot - KIAcc) Passeville, Paul (Crewman - KIAcc)	CAM Bizerte CAM Bizerte CAM Bizerte	Blimp AT-8

Cause	Place	Notes
	Zembra	
-	Gabès (Morocco)	Both Lamoret and Bellot were pilots, Lamoret was in the observer's seat.
Went lost in desert	150 km W. of Nalout (Tunisia)	Wreck of the plane found in the desert on January 10, 1917 Body of Le Bœuf found on January 5, 1918, and remains of de Chatenay on February 17, 1919
Obs ejected during a spin - killed by propeller	Foum Tataouine (Tunisia)	
Lost at sea.	Bizerte / Bône	
	Bizerte	
Falled during bombing training	Bizerte	
	Casablanca (Morocco)	
Accident during reconnaissance flight	Lake of Bizerte	
Fell during training fight	Lake of Bizerte	
	Meknès (Morocco)	
Lost at sea	Off Oran	
Accident on landing	Biskra (Algeria)	
Lost at sea	Off Bône	
	Meknes (Morocco)	
Killed during ground fighting	Aïn-Guettara (Algeria)	
Lost at sea.	Off Marsala	
Accident during reconnaissance flight	Sousse	
	Rabat (Morocco)	
Lost at sea	Off Algiers	Body found on 02-11-1918
Accident during reconnaissance flight	Bizerte	
Accident	3 milles from Cape Serrat	

Attacks on U-Boats from Units of the Aviation Maritime 1914–1918

Date	Unit	Plane	Crew
4/29/17	CAM Bizerte	?	?
4/30/17	CAM Alger	?	?
5/7/17	CAM Alger	Donnet-Denhaut 150 hp Hispano n°261 "A 7"	Cal Nègre, Henri + QM Capdegrelle, Médrard
		Donnet-Denhaut 150 hp Hispano n°251 "A 8"	S/Lt Corren, Léopold + EV 1 Darchis de Lantier, Charles
6/10/17	CAM Oran	?	?
6/28/17	CAM Bizerte	Blimp	
8/9/17	CAM Bizerte	Tellier 200 hp Hispano n°5	?
		Tellier 200 hp Hispano n°79	?
9/19/17	CAM Sousse	?	?
10/10/17	CAM Bone	?	?
10/26/17	CAM Oran	?	?
		?	?
12/5/17	CAM Bone	?	SM Guérin, Robert + EV2 Richer, Robert
4/23/18	CAM Sousse	?	?
		?	?
5/18/18	CAM Arzew	Donnet-Denhaut	EV 2 Jean Leray + QM Henri Baconin
		Donnet-Denhaut	Equipage inconnu
6/9/18	CAM Bizerte (PC Kelibia)	?	?
		?	?
7/8/18	CAM Bizerte	Tellier 200 hp Hispano n°79	?
		Tellier 200 hp Hispano n°80	?
7/8/18	CAM Marsala	?	SM Corouge, André
		?	Cavelier de Cuverville, Albert
8/15/18	CAM Bizerte (PC Kelibia)	?	?
		?	?
9/5/18	CAM Aveiro	Donnet-Denhaut	?
		Donnet-Denhaut	
9/10/18	CAM Bizerte	Donnet-Denhaut n°629	EV1 Henri de Beaumont + SM Marcel Vasson
		Tellier-Canon n°59 "Z-14"	QM Charles Caisson + QM Tanguy Kerros
9/13/18	CAM Aveiro	?	?
9/28/18	CAM Arzew (PC Oran)	?	?
		?	?
10/6/18	CAM Arzew (PC Oran)	?	?
		?	?
10/25/18	CAM Arzew (PC Oran)	?	?
		?	?

Place	Notes
?	2 bombs in 2 attacks
?	1 bomb.
200 km E. Alger, 40 km off the coast	3 bombs. Légion d'honneur for Coren and Darchis de Lantier, médaille militaire for Nègre and Capdegrelle.
	(JORF 03-06-1917 p.4329)
?	4 bombs.
?	8 bombs.
?	4 bombs.
Off Sfax	2 bombs.
?	1 bomb.
?	2 bombs. Hull of U-Boot claimed as hit by bombs.
?	
?	
Large d'Oran	U 39 damaged - Interned in Carthagena until the end of the war.
12 milles NE Cape Bon	U-Boot already attacked 2 days earlier.
?	
?	
2 milles N. Zembia	
3 milles off Aveiro	4 bombs.
Off La Galite island	
30 milles off Aveiro	2 bombs.
20 milles N. and 50 milles W. Oran	4 bombs.
55 milles N. and 60 milles W. Oran	4 bombs.
?	4 bombs, only 2 explode.

French Units of Aeronautique Militaire in North Africa 1914-1918				
Creation	**Name 1**	**Name 2**		**Country**
Jul-16	F. 301	F. 541	Farman	Tunisia
Dec-16	F. 308	VR. 542		Tunisia
Dec-16	C. 543	VR. 543		Tunisia
Jun-17	C. 544	VR. 544		Tunisia
Dec-16	C. 310	VR. 546	Farman, Caudron G.4	Algeria
Jan-16	F. 306	VR. 547	Farman, Caudron G.4	Algeria
Jun-17		548		Algeria
Jun-17		549		Algeria
Jun-16	302	VR. 551		Morocco
Sep-16	305	F. 552	Farman, Voisin VII	Morocco
Jun-17		F. 553		Morocco
Sep-17		F. 554	Farman	Morocco
Sep-17		VR. 555		Morocco
Sep-17		AR. 556		Morocco
Apr-18		VR. 557		Morocco
Apr-18		VR. 558	Farman	Morocco

Gabès and Foum Tataouine	Imprecation in arabic language
Zarzis (opérationnal in April 1917)	Flamingo flying in front of crescent moon (unproven for WW1)
Sfax (Caudron). Then Gabès (VR).	Black Sphinx with yellow wings in a crescent moon (unproven for WW1)
Madhia (Caudron), then Medenine (VR)	None
Biskra	None, but various personal insignas on planes
Biskra (also Beni Mora), Ouragla	Oiseau serpentaire sur croissant rouge
Oran	Targui warrior head on inverted triangle (unproven for WW1)
Constantine	None.
Casablanca, Fez (July 16), Mrirt (July 18), Itzer (August 18)	Red Swastika in white star (unproven for WW1)
Casablanca, Agadir (April 17), Rabat, Tadla (Jan 18), Bou Denib (April 18), Ksar el Souk (July 18)	Leaping gazelle in a crescent (unproven for WW1)
Meknes	Bust of Fatma in a blue 5-pointed star (unproven for WW1)
Marrakech	Scorpio in red disc or green Fatima's hand in a yellow crescent on blue sky.
Taza	Raptor on rock (unproven for WW1)
Rabat , Bou-Denib (Jan 18)	None
Oujda	None
Tadla	Owl perched in a branch, inside a circle marked 558

French Forces Ranks and Their US Equivalents in 1918			
French Army Ranks	**US Army Ranks**	**French Navy Ranks**	**US Navy Ranks**
Général (Gen)	General	Admiral (Am)	Admiral
Colonel (Col)	Colonel	Capitaine de Vaisseau (CV)	Captain
Lieutenant-Colonel (Lt-col)	Lieutenant-Colonel	Capitaine de Frégate (CF)	Commander
Commandant (Cdt)	Major	Capitaine de Corvette (CC)	Lieutenant Commander
Capitaine (Cne)	Captain	Lieutenant de vaisseau (LV)	Lieutenant
Lieutenant (Lt)	First Lieutenant	Enseigne de Vaisseau de 1ere Classe (EV1)	Lieutenant, junior grade
Sous-Lieutenant (S-Lt)	Second Lieutenant	Enseigne de Vaisseau de 2eme Classe (EV2)	Ensign
Aspirant (Asp)	Cadet	Aspirant (Asp)	Plebe
Adjudant-Chef (AdC)		Maître Principal (MP)	Chief Warrant Officer
Adjudant (Adj)		Premier Maître (PM)	Warrant Officer
Sergent-Major (Sgt-Mj)	Regimental Sergeant Major		
Sergent-Chef (Sgt-Ch)	First Sergeant	Maître (Mt)	Chief Petty Officer
Sergent (Sgt) - Maréchal des Logis (MdL) for cavalry and artillery	Sergeant	Second Maître (SM)	Petty Officer 1st or 2nd Class
Caporal (Cal) - Brigadier (Brig), for cavalry and artillery	Corporal	Quartier-Maître (QM)	Petty Officer 3rd class
Soldat de 1ere Classe (Sol)	Private 1st Class	Matelot	Seaman
Soldat de 2eme Classe (Sol)	Private		

Below: French machine gun cars, used to secure the Tunisian border in combination with airplanes, during year 1918.

Made in the USA
Monee, IL
12 September 2026

62974533R00148